U. S. Copyright Office
Copyright July 10, 2017
All Rights Reserved
Richard Miniere
Registration Number
TXu 2-062-083

The True Nature of
TIME
The Biological Theory of Time

Richard Miniere

cover design by Richard Miniere

THIS BOOK IS
DEDICATED TO MY PARENTS

JACK & LORRAINE MINIERE

Quotes About Time

*Time is free,
but it's priceless.*

*You can't own it,
but you can use it.*

*You can't keep it,
but you can spend it.*

*Once you've lost it,
you can never get it back.*

Harvey MaCay

*Time flies like an arrow;
Fruit flies like a banana.*

Anthony G. Oettinger

God's Time is Best.

African Proverb

CHAPTERS

Chapter 1. The True Nature of Time......pg. 1

Chapter 2. Time Does Not Exist in
The Non - Living Universe ....pg. 8

Chapter 3. Time Does Exist in the
Living Universe ....pg. 15

Chapter 4. The History of Our
Understanding of Time ....pg. 25

Chapter 5. Physics and Time ....pg. 30

    Relativity and Time..........pg. 30
    Time – Dilation.................pg. 30
    Dimensions.......................pg. 32
    Parallel Dimensions..........pg. 36
    Space Time.......................pg. 53
    Change and Motion..........pg. 55
    Time and Change.............pg. 56
    What is a Clock................pg. 64
    The Arrow of Time..........pg. 67
    Time Travel......................pg. 69
    Is Time Cyclic or Linear...pg. 72

Chapter 6. The History of the Calendar...pg. 74

Chapter 7. A slightly New Calendar........pg. 87
    Continued on next page

CHAPTERS

Chapter 8. A new Timeline for History.......pg. 104

Chapter 9. Pondering Calendar Reformers..pg. 115

Chapter 10. Lessons about Time..................pg. 127

Diagram # 1. The 28 Day Month................pg. 90

Diagram # 2. The Zero Year and The Zero Century...pg. 110

Books I've Read About Time......................pg. 139

CHAPTER 1

THE TRUE NATURE OF TIME

Time does not exist, except in our minds. That in a nutshell is the true nature of Time. The vast majority of the universe is made up of non-living matter. In this vast, non-living universe, Time does not exist in any way, shape, or form. There is no *past*, there is no *present*, and there is no *future*. The only thing left is *existence*.

Matter, including us, simply *exists* in a *constant state of change*. Except, of course, in that faint whisper of remaining universal matter that we call *life*. Living beings, or more specifically, their minds, are the true domain of Time. Time is basically an evolved, biological, internal guidance system that has been an integral part of our minds since long before we were human. I believe this *Biological Time* is such a basic part of living beings that it probably came into existence either very soon after, or at the same time as life, itself, came into being. I believe this would also be true of other extraterrestrial life that probably exists in the universe.

I believe that Time began when the first life in the universe was born and will end when the last life in the universe dies. We have always thought of Time as a fundamental property of the entire universe. A basic *something*, that caused all movement and change in the universe to occur. Time regulated, with exact precision, the steady, constant flow of the universe from the past, through the present, and on into the future.

What we have always thought about Time is not true. In fact, the opposite is true. Time has absolutely no effect on the inanimate universe because it doesn't exist in the inanimate, or non-living universe. Of course, we all know there is constant change occurring in the universe. The wind blows the trees and bushes around our houses. The river flows, the rooster crows, the farmer hoes, we wiggle our toes, the traffic goes, constant, constant change. All this change we see around us began with, and continues today because of the initial impact of the big bang combined with the physical laws of nature. The momentum imparted by the big bang, along with the physical laws of nature, are the only things needed to keep the universe changing. Time is not needed at all to keep things changing in the non-living universe.

Things are very different in the *Living Universe*. Living things have many needs to remain alive. If those needs are not met, we will die, and most of us have a deep desire to remain alive.

Our basic needs in order of importance are:

BIRTHPLACE.
We need a birthplace. We couldn't exist at all without a planet and a biological environment to exist in.

OXYGEN.

We need oxygen. We can't live more than a few minutes without oxygen.

WATER.
We need water. We can't live more than a week without water.

FOOD.
We need food. We can't live more than a few weeks without food.

SHELTER.
We need shelter. How long we could live without shelter is debatable, but there is no doubt that we need shelter from the elements to survive.

REPRODUCE.
We need to reproduce so we can continue the existence of our species.

These are just the *basic* things we need to survive as individuals and as a species. We need to provide all of these things, all of the time, or we will soon die.

When life began on the earth, billions of years ago, the earth was a naked, watery, and inhospitable place. As soon as our most ancient ancestors came into being on the earth, they were faced with the task of providing for themselves, the basic necessities of life, on a mostly barren planet. I imagine it would have been a lot harder to do under those conditions. The earth was then, and is today, a mostly inanimate object following the laws of physics only, not the laws or needs of life.

The earth didn't need life for its existence, but life did need the earth for its existence.

So, for life to exist it needed to know its home planet exceedingly well. Life needed to know the rhythms and the contours of the earth. Life needed to know the advantages and disadvantages of each. Life needed to know when the sun appeared and how to take advantage of it, and then what to do when it disappeared and how to take advantage of that. The puzzle of constantly predictable rhythms of day, night, month, season and year, mixed with the unpredictability of storm and flood and tide and famine and who knows what else, presented early life with the huge problem of how to cope with all the confusion. I believe that life's answer to this problem was the biological mechanism we call Time.

Now let's compare what living beings need to exist, to what a non-living something needs to exist, such as a rock. What does a rock need to survive? Well, it needs a clump of atoms and-or molecules, and the natural, physical laws to hold its atoms and molecules together, and that's about it. That's all it needs to exist as a non-living rock for billions of years. It doesn't even need to be on a planet. It can spend its entire existence floating through space or buried deep within the bowels of a planet somewhere. It doesn't need to eat or drink or reproduce itself. It doesn't need to strive or struggle to maintain its existence. To remain as it is, it doesn't have to do anything at all.

Now I ask you, which of these two existing things, the non-living rock or the living being, would have any use for Time?

I believe that to exist, the living being would have far more use for Time than a rock, and for the rock to exist, it would have no need for Time at all.

I believe Time evolved in the minds of the first living beings as a biological, internal, navigation system. This allowed living beings to make sense of, and navigate through, the often-confusing environment, external to its own mind and body.

A living being has lots of experiences during its life and our biological internal guidance system, Time, evolved to make sense of these experiences stored in our memories.

We call this store of experiences stored in our memories *The Past*. Time also decided to place the experiences that we are in the process of experiencing in a place we call *The Present*.

The minds of the first living beings began to notice that experiences continued to accumulate as long as the life of a living being continued. Then their minds began to detect a pattern and began to expect all these re-occurring experiences to occur again, even though they hadn't occurred yet. This expectation of un-happened occurrences we call *The Future*.

Time is really our minds master archivist. It causes our memories to be stored in the order of their occurrence so that we can examine our experiences after they have occurred, and learn from them the best course of action to take when, or if, the same experience occurs again. This way we can learn to avoid experiences that could end or severely shorten our lives (lions, tigers, bears), or to seek out experiences that could lengthen or enrich our lives (avoid lions, tigers, bears).

Why would a rock need to know that? Since a rock can neither experience anything learn anything or be aware of anything, what possible use could it have for Time? Since the laws of physics and the momentum imparted to the original atoms of the rock by the big bang, account for its entire existence, why would we need to include a totally unneeded phenomenon to explain the existence of a rock, a mountain, a planet, or a galaxy?

We've all heard of what scientists call our *Biological Clock*.
They talk of this clock being closely attuned to the *Real Time*. They say that the *Real Time*, being that *Universal Property of the Universe,* is what governs the motions of everything, from galaxies to gravestones.

I believe that the *Universal Property of the Universe* type of Time has never existed. I believe that our *Biological Clocks* are the only *Real Time* that has ever been needed or has ever existed throughout the universe. I can sum up the whole thing with a simple equation, $T-M=0$. Time minus the Mind equals Zero.

Since I believe that Time is a wholly biological thing, I've decided to call this theory of mine, *The Theory of Biological Time*.

I've written, "I believe" many times so far in this first chapter, and yet offered no evidence or proof for what I believe to be true.

I've decided to do this, in the first chapter, to explain my theory as simply and clearly as possible. Then go into detail about the whole thing in the remaining chapters of the book.

Going into a lot of detail first thing would have made the book harder to read. Although I would like this book to be a scholarly work, we must also remember that it's subject is near and dear to the hearts of every human being on the planet.

I believe that understanding what Time is would be much more important to people than the discovery of a new sub-atomic particle, or a newly discovered subterranean insect.

In the modern world, our lives are burdened every day by the constraints of Time.

Every person should be able to read and understand what Time really is and why it affects our lives so much. That's why I've written it in a popular style.

CHAPTER 2

TIME DOES NOT EXIST IN THE NON-LIVING UNIVERSE

In Chapter 2, let's discuss the existence of Time in the non-living universe. In Chapter 1, I made the claim that Time does not exist in the non-living universe in any way, shape, or form. I said that there was only *existence* in a constant state of change. We usually think of *Time and Change* as pretty much the same thing. But we also think of Time as being what causes change to occur. However, if we look closely at the situation, we will see that the opposite is true. Time only exists to help living beings contend with all the change occurring around them.

Change occurred first in the universe. The Big Bang is undoubtedly the biggest change that has ever occurred in the universe that we are aware of, and this change has continued to occur ever since. It was billions of years after the big bang that life began to occur in the universe. So, we can see that change was in existence when life wasn't. But did Time exist alongside change?

It's my contention that Time is totally biological. If that's true, then Time and change are separate things, because change was around for billions of years before Time, along with life, came into existence.

To understand this better, we need to define precisely what change is. This definition of change must describe all change in the universe that is possible to occur during the existence of the universe. When we think of all the changes that occur to us in a lifetime, and multiply that by billions of people and all animals, planets, stars, galaxies, and just plain old debris floating around in space, it seems an impossible task. However, it just seems that way.

If we can discover a common denominator among all the change that occurs, we could make some headway.

What could all these changes possibly have in common with each other? The answer to this is on the atomic and sub-atomic scale.

Atoms are always in motion. Even the atoms in solid objects, like a rock, or a steel anvil. Rocks and steel anvils seem very solid and un-movable to us, but on the atomic size level, their atoms are constantly quivering, attached to each other by their constantly moving electrons. If heated until liquid, these same atoms become much more loosely connected to each other, and move around even faster, bumping into each other in a totally random motion. If heated to a gas, the atoms become completely un-attached to each other, and really fly around with great energy and spread out in every direction, unless captured in a container.

Robert Brown first detected this motion of the atoms in the early 1800's. He was a Scottish botanist who first discovered the nucleus in living cells. He also noticed that some unknown force was constantly jiggling around pollen grains in water, under high magnification. At first, being a botanist, he thought that some mysterious *life force* in the pollen was causing the pollen grains to jiggle around. But when he looked at non-living dye particles, he saw the dye particles being bumped around in the same way by the random motion of the water molecules.

He had no idea what was causing it, but he did report it, and in his honor this motion is today called "Brownian Motion". It was the first observed physical evidence for the existence of atoms. Since everything physical is made up of billions of these atoms clumped together in various ways, I think that this *"Brownian Motion"* would make an excellent common denominator of change.

Now let's dissect just one atom. The simplest atom is the hydrogen atom, so we'll start with it. The best way to see this atom is to make a model of it. Since atoms are so incredibly small, we'll have to make an incredibly large model just to see it. The nucleus of the hydrogen atom contains two particles, the neutron and the proton. They rotate closely around each other, and in our model, they, both together, are about the size of a standard baseball. That doesn't seem very large, but to be a complete atom, it has to have at least one electron in orbit around it, to connect it to other atoms.

In our super large model, the electron is about the size of a grain of sand rotating around the baseball at a precise distance from the nucleus, or baseball, of about two and one half miles. This would make the diameter of our model hydrogen atom 5 miles across. The electron rotates around the nucleus in a random order, like when you roll up a ball of string. It rotates around the nucleus billions of times per second in our model. The electron would seem to be everywhere at once, like the spinning blades of an airplane propeller, making it look like a solid sphere.

What causes these electrons to constantly and reliably rotate so fast is completely unknown. Brownian Motion and the motion of electrons is what we'll use as our common denominator of change. Since these particles are constantly changing position in relation to each other, I'll call this common denominator Change of Physical Position, or COPP. It works on a larger scale too. Two boys playing catch with a baseball, or the moon swinging around the earth, or two galaxies rotating around each other, are other examples of *"Change of Physical Position"*.

Our next question is: how does *Time* relate to this *Change of Physical Position*? Well, the 20th century idea of Time as a Universal Property of the Universe is an attempt to explain exactly the same thing as the *"Change of Physical Position"* explanation does. So, we can say that in any process, formula, sentence, physical, or mathematical construction, the TIME explanation = the COPP explanation. Since we know that physical things exist, and that physical things constantly change position in relation to each other, we know that COPP is a real, physical thing. But is Time a real physical thing?

I don't believe the physical, 20th century," Universal Property of the Universe" type of Time is real at all, and I'll tell you why. Time is not needed to explain anything in the real, physical world. We can explain it in other, physical ways equally as well. When we measure how much Time it takes to do something, for instance, a man walking a mile, we are not comparing his performance to a mysterious, unknown something. We are comparing his Change of Physical Position to the very regular Change of Physical Position of the sun in the sky, or to a small mechanical meter, a watch, designed to mimic the very regular motion of the sun. We have also un-wittingly designed the watch to mimic the Biological Clock in our heads.

In any chemical formula, or process, Time is not used at all.
Oxygen, for instance, combines with other elements rapidly, as in a steel wool fire, or slowly, as in a rusting iron bar. Both of these chemical reactions are the same thing, and can be written down without using Time in the formula. The steel wool burns rapidly only because it's in the form of fine slivers of steel. Most of the volume of this steel is surrounded by an equal volume of oxygen ignited by intense heat, whereas a steel bar won't burn, but rusts slowly because most of the steel is not exposed to oxygen, just the outside surface, and very little heat. But the process is the same, even though it will take years for the iron bar to rust away completely. As in the example above, it's really just a matter of physical amounts.

Let's look at a second example. If we cut the bottom entirely out of a 5-gallon plastic paint bucket and then pour exactly 4 gallons of water into it, it would drain instantly. But if we drilled a 1-inch hole in the bottom of another 5-gallon plastic paint bucket, and pour exactly 4 gallons of water into it, it would probably take 5 minutes to drain. We humans would describe the difference in terms of Time.

One bucket drained instantly, the other bucket took 5 minutes. However, we can also describe what happened without referring to Time at all. We could say that the water in the bottomless bucket experienced a large amount of Change of Physical Position compared to a small amount of Change of Physical Position of the sun because there was no restriction between the water and gravity. We could then say that the water in the 1 inch hole bucket experienced a very small amount of Change of physical Position compared to a larger amount of Change of Physical Position of the sun, because of the large amount of restriction between the water, and gravity.

    The above is a really clumsy, but accurate, way of saying, "one bucket drained instantly, the other bucket took 5 minutes." But it's a very accurate way of showing that the 2 physical happenings can be described in terms of physical amounts only without referring to Time at all. If we have two explanations that equal the same thing, and one can't be shown to exist, and the other explanation can be shown to exist, then the one that can't be shown to exist should be discarded. I think the old 20th century, "Universal

Property of the Universe" type of Time can't be shown to exist, and should be discarded.

CHAPTER 3

TIME DOES EXIST IN THE LIVING UNIVERSE

As I said in the first chapter, Time only exists in that faint remaining whisper of universal matter that we call life. Time is an evolved, biological archival system that is hard-wired into our minds. It's the Biological Clock that scientists are always talking about. Scientists who study the brain tell us that we seem to have 2 different clocks in our heads. One is an "Interval Timer", a sort of stopwatch. I'm sure that musicians and acrobats are experts at using their mental stopwatches. The other is a Circadian-clock, a sort of daily cycle timer that regulates our waking, eating, daily activities, and finally our sleeping. This is the clock that jet lag throws for a tailspin.

I believe that our internal timekeepers do pretty much the same thing as our external, mechanical, "Time-Keepers" do. They use some internal body process of some kind, a biological "pendulum" of sorts, that constantly changes in a very regular manner to compare to irregular events external to our bodies. Could this be our heart? Our heart beats pretty regularly most of the time, but not always. There is quite a difference between a resting heart rate and a highly-stressed heart rate, so I'm not sure that the heart is the body's pendulum. I suspect there is another chemical or electrical process in the body that is very regular at all times, that our internal, biological clocks use as a pendulum.

In our minds the idea of *Past, Present and Future*, and the idea of *Before and After*, are held with such a firm, iron fisted grip, that it is almost impossible for us to think any other way. It's just the way our minds work. The reason is that it works well, and it is vital to our survival that our minds work this way. To us, and our ancestors, going back to the beginning of life on earth, Time has been as vital to us as food, and it still is today, and will be as long as we live. Evidence that our minds work in a *Past-Present-Future* way is obvious in our spoken and written languages. We all speak different combinations of words, or time-stamps, to get across the idea of when we do something.

For example,

Past - "I WENT fishing."

Present "I AM fishing."

Future "I WILL GO fishing."

There are hundreds of languages all over the earth that all have the same *Past-Present-Future* tenses, using different words that mean the same thing. We use a lot of words that are "time stamps". A very short version of past, present, and future is yesterday, today, and tomorrow. We say other time stamp words like, someday, once upon a time, pretty soon, forever, not long now, once in a blue moon, in the far distant future, in the far distant past, right now, an eternity. These words are good evidence that our minds all work the same way. As you can see, this business of *Past-Present-Future* is very necessary and deeply embedded in our minds.

Now let's look at how Time helps us navigate through our lives. Earlier I said that Time is our *Master Archivist*. If we compare our minds' Time, to other, external examples of archiving, we can see how similar they are. First let's look at books in a library. Imagine a large library as big as a city block that is ten stories tall. Every floor is filled from ceiling to floor with books. Now imagine some roving reprobate breaks into the library at 2am and tosses all the books out the windows into the streets below. Then he takes all the labels off all the shelves, and escapes before dawn. What a mess! You could still read the books but you couldn't find a specific book. It would take months of hard work to bring the library back to its full usefulness, simply because the books were not stored in an orderly manner.

Second, let's look at a modern TV news station. They are constantly recording videos of all kinds of events every day. This recording has gone on for decades. It's hard to tell when an event took place by looking at the video itself, because video images have been very clear for many decades now. Look Out! Here comes our early morning reprobate again. He peels all the dates off all the videos and dumps them into a large pile on the newsroom floor. The videos have not been destroyed, but their order has. Their usefulness has been greatly diminished. It could take months or even years to piece back together the puzzle of the videos correct order.

Third, let's look at archeological excavations. The first archeologists would just dig a big hole in the ground to see how many bones they could find. They would toss the bones they found onto a big pile and examine them later. Eventually they realized that by not recording the positions of the bones in relation to each other and in relation to the vertical, stratified layers of rock and soil where they were buried, they were destroying knowledge of the buried artifacts. They were also destroying any lessons that could be learned from that knowledge.

I'm sure there are dozens of other examples I could give, but these 3 are enough to prove my point that our minds use time to do the same thing as the three examples above do. But whatever words you use, archiving, positional information or an orderly system of retrieval, they all lead to the same conclusion. Our minds need to record our experiences in the real world and then be able to replay and examine those experiences in their proper order of occurrence. That way we can learn from them how to better handle experiences we've had, and haven't had yet.

As I talked about, at the beginning of this chapter, one of the things Time has given us is a certain viewpoint towards our memories.

Memories are simply experiences recorded by our minds. An experience is an interaction between the physical world and our minds. In order for our minds to extract the greatest amount of useful information from our experiences, Time makes sure that our experiences, or memories, are recorded in their proper order of occurrence. Time then directs our minds to store all of the memories of our experiences in the correct order of their occurrence, in a category that we call *The Past*. A sub-category of *The Past* is *Before and After*. Since the past is the only place memories are stored, I think *Before and After* is also a broader kind of mental sorting device.

The Past and Before and After.

My Father had a great memory. He could remember things from when he was a little kid. He told me of when he was in the third grade he had told a friend that his Grandpa had chickens that were so big that they were as high as his shoulder! His friend was pretty skeptical. Dad finally realized that it wasn't that the chickens were so big, it was just that he was so small at the time he remembered them. Dad was just getting used to using his memory in the third grade when you don't have much to remember yet.

My Dad's greatest adventure in life was being a B-24 bomber pilot during WWII. He flew 50 bombing missions in and around New Guinea in the south pacific, just north of Australia, where he won the Distinguished Flying Cross and the Air

Medal. After that, he ferried bombers all over the U.S. and around the world, and he met my Mother, all in the span of 5 years. After that, my folks had me, my Dad went back to work and back to the same dull life he had before he joined the army. When I was about 12, my Dad was about 37. At that time, whenever he told a story about his life, there were three major divisions of time. There was *Before the War, During the War and After the War*.

"Oh, that happened *before* the war", he would say. "I had that old Buick *during* the war". "I didn't see him again till *after* the war". He couldn't remember the exact dates sometimes, but he could remember what came before or after something else. Everybody knew *which* war. I'm sure it was the same all over the world at that time.

The Present or *Now*.

Time also puts the experiences that we are in the process of
experiencing into a category that we call *The Present, or Now*.
What is *Now* anyway? Now is relative. When a bunch of soldiers huddle to synchronize their watches, *Now* means RIGHT NOW! But if we're talking about the Roman Empire, *Now* could mean the last hundred years, or The Present. If we're talking about 100 million years ago during the age of dinosaurs, then *The Present* could mean the age of man, or the last 5 thousand years! *The Present* seems to be a broader, stretched out version of

Now. A pretty vague answer, I'll agree. So what exactly is *Now*?

If we look closely at this question we'll see that *Now*, is about a 30th of a second long. Movie cameras, and later video cameras, take a still picture about 30 times a second. When these pictures are projected at that speed the movie looks like real life. If we project still pictures taken at 30 times a second, at 15 times a second, then our biological vision can detect a bunch of still pictures changing rapidly
So, it's pretty obvious that our brains' vision sees the real world at about 30 pictures a second, or a 30th of a second. I could be wrong, but our other senses are probably recorded into our memories at about the same rate.

While we're on the subject of *Now*, what would it be like if, by some accident, we were trapped in *Now*? This is a silly sounding question, but it actually happened to a man. He is an Englishman named Clive Wearing, and he has the worst case of amnesia ever known. On March 27, 1985, he contracted herpes viral encephalitis, a herpes simplex virus that attacked his central nervous system. Before getting this disease, he was a very accomplished musician and singer in London, England. His accomplishments are far too many to list here, but one of the high points of his career was while working for BBC radio 3, he was responsible for the music played on the BBC during the day of the wedding of Prince Charles and Diana Spencer. This poor guy can't remember the past beyond about 12 seconds ago. He can't form new memories, or recall old ones, and he has very little control over his emotions. When his wife, Debra, visits him in the mental hospital where he lives, he greets her with a big, prolonged hug and says "Oh Debra! I haven't seen you in ages, how have you been"? They would chat for about 12 seconds and then he would say "Oh Debra! It's been so long since I've seen you! This went on continuously while his wife was there. It seemed as if his mental memory recorder was being reset to zero every 12 seconds or so. On YouTube, you can see the TV show where I first saw him on the PBS series *The Mind, episode 1, In Search of the Mind.*

What a horrible fate, being trapped in your own brain with no memory of the past! Most of us are a lot luckier than we think!

The Future

And last of all, over the eons, the minds of living things have learned, through experience, that as long as living beings stay alive, experiences will continue to occur. Time, our biological archivist, then places these Expectations of Continued Occurrences in a category we call *The Future*. We tend to think of *The Future* as something completely un-knowable. *The Future* is beyond our comprehension, while *the Past* is all recorded, written down and known. Is this true? Is *the Past* completely known? Is *the Future* completely unknown? Just by reading the last two sentences I'm sure you're beginning to have your doubts. We all know that there are many unknown things about *the Past*. What happened to Jimmy Hoffa? Who is buried in the tomb of the "unknown soldier"? What happened to Amelia Earhart? There are millions of other questions like this that we have no answer for.

We also know millions of things that will happen in *the Future*.
It will rain. There will be earthquakes. People will kill each other. People will make things. People will get married and have kids. People will be born and die. Animals will be born and die. Birds will fly. Horses will run. Fish will swim. Rivers will flow. Mountains will erode. Clouds will form. Minds will be changed. Dreams will be dreamed. Buildings will be built. Ships will sink.

I believe there is no difference between *the Past* and *the Future* in the real world. I believe there is only existence in a constant state of change. I believe that this *Past-Present-Future* business only exists in our heads, to be used by our minds, to try and comprehend the constant state of change taking place in the world around us.

On a religious note, the Catholic Church believes that God exists *outside* of Time, while only man exists inside of Time. This would be true according to the *Biological Theory of Time* because if God exists and he knows everything then he can't learn anything new. So he would have no use for Time, because Time is a biological learning device. Time would be as useless to God as a gas tank on an electric car.

To understand the true nature of Time we must learn to think outside of our minds, at least for a little while. We shouldn't try to do this permanently of course. To abandon our wonderful, God-given, *Biological Time Mechanism* would be the height of foolishness. Even if we could give it up. The reason we shouldn't abandon it is because it works so well!

If you want to explore the deep mysteries of Time, the brain is the place to explore, not the cosmos.

CHAPTER 4

THE HISTORY OF OUR UNDERSTANDING OF TIME

The history of our understanding of Time begins with the caveman, and progresses up to the time of Isaac Newton, when Time first began to be described scientifically. Then we look at Time from Newton to Einstein, and up to the present day, and the *Theory of Biological Time.*

During the time of cavemen, up till Newton's time, people thought of Time as a sort of natural feeling of things passing through their lives. They couldn't explain what it was exactly, but they knew that it existed and they used it as much as we do today. A perfect person to illustrate this viewpoint was St. Augustine of Hippo. No, he didn't own hippos or anything. Hippo was a town in North Africa where he lived, in today's Libya. St. Augustine was very influential in the early Christian Church. When the Roman Empire collapsed in 410 A.D., the City of Rome was sacked by the Germanic tribes. The people of Rome thought that they were being punished by the Gods for abandoning their traditional Roman Gods in favor of Christianity.

St. Augustine wrote a book to deny this called "The City of God" to assure people that Christianity was not at fault. Another famous book he wrote called "Confessions", told of his sinful youth, and his ultimate salvation by turning to the Christian faith.

St. Augustine famously said about Time,

"When I'm alone, I know what Time is, but when someone asks me to explain it to them, I cannot".

When we in the 21st century read this from the 5th century, 1600 years of Time between him and us are instantly erased. I think every one of us has felt the same way.

Isaac Newton was born about 1300 years after St. Augustine. He became the greatest scientist who ever lived. He invented Calculus, he proved that white light contained all the colors of the rainbow by using a prism, he invented the Reflecting Telescope, he was a mathematical genius and was said to have advanced every branch of mathematics then studied. His most famous work is his book, "Philosophiae Naturalis Principia Mathematica" in Latin, or "Mathematical Principles of Natural Philosophy" in English, or just "Principia", was published in 1687. It deals with Newton's Laws of Motion and Planetary Motion.

In his book, he describes his belief that Time and space are totally separate things. He also believed that there are 2 types of Time, and 2 types of space. There was Absolute space or Time, which was absolutely perfect and immovable, which could only be detected mathematically, by mathematicians! Then there was Relative space or Time, which only applied to the average person when they were looking at the sun, moon or stars, and moving things on earth to tell time. This belief is often referred to as Newtonian Time or Space. This sounds like something a mathematician would dream up. They're always working with absolute perfection. Perfect spheres, perfect triangles, perfect lines, perfect angles, perfect planes, perfect things that don't exist.

Gottfried Leibniz and Immanuel Kant, 2 other famous scientists of that time, held an opposing view that time was a fundamental intellectual exercise used by humans to compare to events to tell how long the event lasted. They thought that Time or space was not an event, or a physical thing. Pretty close!

Then, About 2 hundred years later, in 1905, 25 year- old Albert Einstein, while working as a patent clerk, wrote 4 papers that shocked the scientific world. 1905 was called his Miracle Year.

The papers were about:

The Photo Electric Effect
   He said that energy came in small pieces called quanta.

Brownian Motion
   He said that Brownian motion was real evidence for Atomic Theory

Special Relativity
   He said that the speed of light is independent of the motion of the observer, and that there is no light bearing, or luminiferous, ether that carries light waves.

Matter equals energy

He said that Matter is concentrated energy, E=mc/2. Energy equals Mass x the speed of light, squared.

This was not the end of his productivity, but it was a remarkable beginning. The Physical sciences were never the same again. One of the things he proposed was that the 3 dimensions of space, and the 1 dimension of Time should be combined into a 4 dimensional something he called *Space-Time*. In other words, if you want to meet somebody in a large skyscraper you would need to know their 3-dimensional location (floor, room number), and it's 1 dimensional time location, to meet them. Einstein also said that according to Relativity, Time was not rigid but flexible, and would stretch out a great deal under certain conditions.

Now, we come to the Theory of Biological Time. What does this new theory have to say about *Time*? According to the Theory of Biological Time, the only Time that exists is imprisoned within the bodies of living beings. So, that would mean that Time does not exist in the non-living universe as a whole. However, if we remember that *Change of Physical Position* is what we have always thought Time to be, then we can see that *Space-Time* is actually *Space-Copp*.

Now, let's look at space itself. Space is really just total nothingness. However, space is also filled with all the matter in the universe. This matter is spread all through the universe, but not evenly. It's concentrated, mostly in separate clumps called galaxies, stars, planets and gas clouds with vast, comparatively empty spaces between them. We can't measure space because there's nothing there to measure. Remember that when we measure something we are comparing a known physical amount to an unknown physical amount, for instance, we compare a ruler of known physical length, to a physical board of unknown length. However, the clumps of matter inhabiting space are measurable. We can measure the width and the depth of the clumps, we can measure the number of the separate clumps, and we can measure the distances between the clumps, but we can't measure anything if there are no clumps, or gas, or sub atomic particles there to measure.

So, we can see that a better name for *Space-Copp* would be *Matter-Copp* because matter and its constantly reliable change of physical position are real, and can be measured, but *Space*, and *Time*, external to our bodies, are not real and cannot be measured. *Space-Time* is supposed to be a combination of 2 different aspects of the same thing, but neither of them exists. However, matter and its constant motion are different aspects of the same thing. So, I think we should replace *Space-Time* with *Matter-Copp*. So, *Matter-Copp* it is.

It's remarkable that Einstein combined *Space* and *Time*, even though he didn't really know exactly what time was. He was right to do so because *Matter* and its constant *Change of Physical Position* perfectly fits into the hole left when we remove *Space-Time* from his theory

CHAPTER 5

PHYSICS AND TIME

RELATIVITY AND TIME

Albert Einstein's theory of relativity has a lot to say about Time. He took Time from a rigid background sub-structure to which all occurrences were firmly fixed, to a flexible, stretchable background fabric where Time could be stretched, pulled and deformed according to certain physical conditions.

TIME DILATION

Albert Einstein's Theory of Relativity tells us that nothing can move faster than the speed of light. The speed of light is a universal speed limit. If we were riding in a spaceship that was moving at 10% the speed of light, we would be using a certain amount of fuel to propel it. If we advanced the throttle to 50% of the speed of light, the mass of the ship would increase greatly, as would the amount of fuel consumed. If we reached 99.9% of the speed of light, the mass of the ship would be near infinity, and so would the amount of fuel needed to get it to that speed, but of course we would run out of fuel long before that.

The theory of relativity also tells us that as we approach the speed of light, the length of the spacecraft would be compressed in the direction it was traveling. The faster it went the shorter it would get until when the ship reached almost the speed of light, the nose to tail dimension would almost be flat. This would happen to the pilots, their equipment, their supplies, any clocks onboard, and even to any motion, or *Change of Physical Position* on the ship. Since the *Change of Physical Position* would be compressed, every motion on board would slow down and almost stop near the speed of light. A minute on board the speeding spaceship would seem like a minute to the pilots, but it would be like a month or a year to us on earth. This slowing, or stretching out of Time, is what Einstein called *Time Dilation*, and it's been proven to be true.

    Although the brain is made of living tissue, the single atoms that make up the brain are not living, and are subject to the same physical laws as any other atomic matter. That's why the real Time in the pilot's brain and the mechanical Time of the clock are both affected equally by *Time Dilation,* or *Change of Physical Position Dilation.*

    Actually, I think it would be more accurate to call it *Change of Physical Position Compression* because the ship traveling near the speed of light would have its normal change of physical position compressed down to almost nothing, while back on earth a normal amount of change of physical position would have taken place.

DIMENSIONS

The 3 dimensions of space originated with the Greek mathematician Euclid, who was the Father of Geometry. His 3-coordinate system of measuring any possible point in space, or to enclose any amount of space, uses 3 directions that are 90 degrees from each other. They are usually labeled X, Y, and Z, or commonly as length, width, and height. But first, we have to have a starting point. The starting point is on a flat 2-dimensional plane. These 3 numbers are thought of as the bare minimum number of dimensions needed to define any point in space, or to enclose any object in a certain amount of space. This was the norm for a couple of thousand years. People had long accepted the idea of 3 basic dimensions and never questioned the idea.

Then, Albert Einstein came along and decided to add a 4th dimension of *Time* to the mix. Since *Time* and *Space* seemed to have nothing in common with each other at all, he decided to call this 4th dimensional construct *The Space-Time Continuum*. I guess the idea was that the word continuum was a sort of verbal glue to convince everybody that these 2 wildly different things were actually one thing.

Nobody had ever suggested such a thing before, and this got people to wondering if there could be even more than 4 dimensions, maybe 5, or 6, or more?

This idea soon popped up in popular culture in, of all places, the Superman comics. One day a

Mr. Mxyzptlk, who supposedly came from the 5th dimension, appeared in the Superman comic strip. He was a trickster, more of a nuisance than a danger, who delighted in annoying Superman. Superman somehow discovered that he could send Mr. Mxyzptlk packing back to the 5th dimension if he could only trick Mr. Mxyzptlk into pronouncing his own name backwards (the joke being that his name was pretty much unpronounceable either way). Future physicists (of today) seem to have cut their teeth on this idea, so that today, they seem to think that we need at least 11 basic dimensions to explain the universe. They also seem to think that there are multiple universes out there somewhere. Some even say an unlimited number of universes.

Where is Occam's Razor when you need it? Occam's Razor is a problem solving principle invented by William of Occam (1287-1347), an English Franciscan friar, philosopher, and theologian. His principle states that among 2 or more competing theories, the simplest explanation should be chosen because simpler theories are more easily tested. I guess they called it Occam's Razor because it cut through the crap so well.

Are there really eleven basic dimensions? Are there Multiple Universes, or an unlimited number of Universes?

I don't pretend to have any idea what they are talking about, nor do I pretend to think they have any idea what they're talking about.
All I know is dimensions are just measurements. Let's look at a 747 airliner. It has 3 basic dimensions of height, width, and depth, which

would describe a big box that would envelope the airplane. But there are millions of other dimensions (measurements) to describe the whole airplane. All of which are just as valid dimensions as the first 3 are. Why are the first 3 considered basic and almost magical? I've thought about this many times over the years and I've come to believe that the emphasis on the 3 basic dimensions came about because of the architectural and building trades.

Humans have been drawing and building buildings for many thousands of years now, and by far the easiest way to build a building is to build it square. We can build round buildings, but it's much more difficult to do. They're usually restricted to churches, government buildings and stadiums, but the vast majority are built square all over the world. The reasons are simple. A square is easier to draw. A square is easier to measure. A square is easier to add on to. A square is easier to layout on the ground. A square is easier to build, because it's easier to check the accuracy of your work with simple tools. A string pulled taut to the edge of a wall being built, gives instant feedback to accuracy. So does line of sight without any tools at all. A small stone tied to the end of a string and hung vertically against a wall will show if the wall is properly vertical or not. Another simple, quick check of a square or rectangular building is to run a twine between the 2 diagonal corners of the square. If the 2 measurements are the same, then the building is square. None of these measurements would work as well on a round building, except for the vertical measurement. All ancient builders knew about these simple tools and knew how to use them. When you consider that many great thinkers of the past were also mathematicians and architects (Leonardo Da Vinci for instance), we can see why they might assume that there were 3 basic dimensions. Modern builders still use these same 3 basic dimensions.

Let's think about these 3 basic dimensions some more. The 3 basic dimensions describe a cube. Cubes occur in nature as crystals as in salt crystals, but they don't occur that often. By far the most common physical shape that occurs in nature, and in the universe, is the sphere. Every star and every planet is a sphere. Even galaxies are round if not spherical. Then, when we plunge down into the depths of the atomic level of size, we see that atoms are also spherical. I don't think I'd be too far off base to suggest that the sphere is the basic shape of the universe. And how many basic Dimensions does a sphere have? It only has 1 dimension, its diameter, or radius. As stated at the beginning of this article, the 3 basic dimensions are needed to define any point in space, or to enclose any amount of space. A one dimensional, imaginary sphere can enclose any amount of space as well as an imaginary cube, and you only need one dimension to do it, the diameter, or radius.

As for defining a point in space, you really only need to define 2 points in space, the point that you're at, and the point you want to measure to. I think it would be easier to just measure a straight-line measurement between any 2 objects. After all, that's what they did on the Voyager spacecraft. They put a drawing on Voyager to show straight-line measurements between Earth and various Quasars, so if the spacecraft were found by aliens they would know where to find us. All you'd need would be 2 quasars to triangulate where earth is located. I don't know if that was such a good idea? Who knows who's cruising around out there?

So maybe the basic 3 dimensions are not as basic, or sacred, as we thought.

PARALLEL DIMENSIONS

Scientists have suggested that there are, what they call, Parallel Dimensions that exist, but we can't see or comprehend them, except through mathematics. Those mathematicians have all the luck! These Parallel Dimensions are supposed to be right beside us but we can't see or feel them, sort of like a room next door that we can't see or detect, not even by banging on the wall! Maybe if we banged on the wall with an advanced math textbook?

In order to enlighten us to this situation, they give us the example of a poor one-dimensional creature that is completely unaware of the "higher" dimensions of 2 or more. This supposed creature is simply a straight line between 2 points. It ought to have a name. Let's call it "One-zee." Now the next creature up has the 2 dimensions of length and width, but zero height. This creature is as flat as your wallet on the day before payday! Let's call this higher creature "Two-zee". Two-zee is totally aware of itself, and One-zee, but poor One-zee is clueless about anything but itself. The 3rd dimension has length, width, and height, so to remain consistent we'll name this creature "Three-zee". Three-zee is also aware of itself and all lower dimensions, however there is no motion in the 3rd dimension because it has no *Time*. Think of a non-moving, lethargic teenager draped over a chair, trying to figure out how to avoid doing his homework, and get away with it (I've been there), and you'll have a good idea what a Three-zee creature might look like. Now let's look at the dimension that you and I inhabit, the 4th dimension. Yes, you and I are "Four-zees" because we have 3 dimensions and can move around. We Four-zees are all aware of the 3 lower, and our own, dimensions, but unlike those lower dimensional creatures, thanks to the research done by Superman comics, we are aware of the 5th dimension, and by the research done by future mathematicians, who read those Superman comics as kids, we are also aware of the 6th dimension and still higher dimensions on up to the 11th dimension.

    Mathematicians and physicists also seem to think that there are multiple universes that are parallel to our own. I'm not sure what the difference is between a parallel universe, and a parallel dimension, or are they the same thing? Who knows! I am not a theoretical physicist or a theoretical mathematician, so I don't pretend to know what they mean when they talk about parallel, or alternate, dimensions that are supposed to exist side by side without being aware of each other's existence.

However, I do know what a dimension is. A dimension is simply a measurement of physical matter. Anything that can be measured has a dimension. I do believe a case can be made for the existence of parallel dimensions because I live in one, my house. I've lived here for about 6 years and there is a parallel dimension that exists along side me here that appears to me at different times. They're called ants. I'm not talking about normal sized ants, these teeny, tiny things are about the length, but not the width, of a lower case, typewritten "o". It's amazing that anything that small can be alive! I'm sure there's a whole population of them up there in the attic somewhere that's been there for 45 years since the house was first built. To them, the house is their hollow mountain where they inhabit the "upper caverns". There's a giant (me) living in the "lower cavern" but he's too big to evict, so they don't worry about it. Occasionally, when they do appear, they usually show up on the top of my kitchen counter tops.

Once I noticed a spot on my counter top about the size of the "o" at the beginning of this sentence. It looked like a miniature sun with little rays around it! Looking closer I saw a herd of these micro ants chowing down on a drop of gravy. I'm always amazed at how small they are. I held my hand out about an inch above the counter top and waved it back and forth over them. There was absolutely no reaction from them at all. They paid about as much attention to my passing hand as a group of humans on a picnic would pay to a passing cloud. Another time, when they had found another micro snack, I found them in a line crawling across the full length of my counter top. I wondered where they had come from, so I followed them to see. When they got to the wall, they went up it to the ceiling where they crawled along the ceiling about 3 feet from the wall, and just disappeared. I got a ladder and a flashlight and got my face about 6 inches from the ceiling where they were disappearing, and shined the flashlight there. There was a chip of paint that was so small that I never noticed it before while standing on the floor, and the ants were squeezing through it to get into the attic. Absolutely amazing, yet it happens every day somewhere. The only way I've ever gotten any reaction from them was when I sprayed bug spray on them, and I'm sure they had no idea where it came from.

In this example, at least one of us knew of the other's existence. There are many examples where neither party are aware of each other's existence. Before the microscope was invented, people were totally unaware that there were great populations of microscopic creatures living in their guts that were necessary for them to digest their food. This was a perfect example of parallel dimensions, where neither of the beings, the humans or the intestinal microbes, were aware of the other's existence. Let's look at this further by looking at the largest living things on down to the smallest living things, and their relationships to each other.

First, let's look at the largest living things on planet earth. The largest living things on earth are plants. The largest living plants are things called "Clonal Colonies". These are plants that sprout, grow (or clone) when adult, and send out horizontal root systems, which then sprout new vertical plants, or stems. Eventually the original plant even dies, but the root system never does, it just keeps on sprouting new stems, and growing the root system for thousands of years. The largest and oldest of these clonal colonies is called "Neptune Sea Grass", which is native to the western Mediterranean Sea near the island of Ibiza, which is just off the eastern coast of Spain. In 2006, a Neptune Sea Grass colony there was discovered to be 8 kilometers, or 5 miles across, and estimated to be 100,000 years old. WOW!

Another huge clonal colony is the Armillaria Solidipes Fungus. It lives on the sap of hardwoods and conifers west of the Cascade mountain range in the Pacific northwest of Oregon, in the United States. This fungus can travel great distances under tree bark and underground by using a system of thin roots called "shoestrings". A single specimen found in Malheur National Forest in Oregon is thought to have been growing for 2,400 years. Today it covers an area of 3.4 square miles or 2,176 acres. Most of it is invisible underground. The locals call it the "Humungous Fungus."

Still another clonal plant is the "King Clone" Creosote plant. The creosote plant is just a bush about 3 feet tall that uses the clonal method to reproduce. The original plant sends out lateral roots that sprout and grow new plants around itself. When the original plant dies, it decays away and leaves a hole in the center of the children plants forming a ring of bushes. When the children plants die, the ring gets larger, and the size of the ring grows. However, the plant is still all the same plant. The "King Clone" Creosote bush has been alive for about 11,700 years and has reached a diameter of 67 feet. It lives in the "Creosote Rings Preserve" of the Lucerne Valley in the central Mojave Desert in California.

And finally, there's Pando, which means "I spread" in Latin. Pando is a clonal colony of Quaking Aspen trees living in Utah about 1 mile southwest of Fish Lake on Utah's Route 25.

Pando was discovered in 1968 by Burton V. Barnes who identified it as a single genetic organism. Years later, in 1992, Michael Grant of the University of Colorado at Boulder looked at this clonal colony, named it Pando, and claimed it's trees and root system to be the world's most massive organism. Instead of its "stems" being small bushes, grass, or toadstools, Pando's "stems" are entire Aspen trees all connected by one gigantic root system. One "stem" is about 80 feet tall and can live for 120-150 years. Eventually, each tree dies, rots or is burned, but the root system lives on and later sprouts new "stems", or trees. Pando covers 106 acres, or about 7,000,000 square feet, weighs 6000 tons, or 13,000,000 pounds, and is thought to be 80,000 years old. I'm sure that anyone walking through this grove of trees in the 1950's and before, were totally unaware of this Parallel Dimension.

The next smaller living things are still plants. They are individual trees. The oldest living individual tree is the bristlecone pine. The oldest living example is about 5000 years old. The most massive individual tree, and also the most massive living, individual organism, is the Giant Sequoia Redwood of California, which lives about 3500 years. The General Sherman giant redwood is the largest known, living, single stem tree. It's 275 feet tall, has a volume of 52,508 cubic feet, and weighs 2100 tons, or 4,200,000 pounds. Insects living on this tree must think of it as their own planet, if they do any thinking at all.

Now, let's go to a little smaller dimension to the largest living animals. The Blue Whale is the world's largest living animal, and also the most massive animal that has ever lived on earth. It's about 100 feet long and weighs about 190 tons. That's about 491,000 pounds.
It's amazing that the world's largest animal feeds on one of the world's smallest animals, called krill, which are like tiny shrimp, only about a half inch long. This is a great example of parallel dimensions. Of course the whale knows that krill exist, but do krill know the whale exists? They do when it's too late I guess. It's also interesting that krill are also filter feeders just like the blue whale is, but feeding instead on microscopic plankton. There are many species of krill around the world, but the Blue Whale likes to eat the Antarctic Krill,

which is estimated to have a biomass of 379,000,000 tons. The Blue Whale can eat about 7,900 pounds of Krill a day. I wonder if the whale even knows that the krill it eats is even alive? Maybe the whale just thinks its food, like when a human eats a bowl of grits or cream of wheat. The whale and the human know their meal is made up of tiny particles, but all each cares about is that it's good to eat.

The largest non-whale fish in the world is the Whale shark. This shark has 300 to 350 rows of tiny teeth, and 10 filter pads to filter feed just like the Blue Whale. The largest verified specimen was 41 feet long and weighed 47,000 pounds. Most whale sharks are smaller than this, but there are reports of some that are bigger.

Now let's move on to the biggest land animals. The biggest land animals that ever lived on earth were the Dinosaurs, the largest of which were many times larger and heavier than the largest land animal alive today, the African Elephant. But that's not to say the African Elephant is a runt. The African Bush Elephant stands about 10 to 13 feet high at the shoulder and weighs about 10,000 to 13,000 pounds, or 6 and one half tons. The Giraffe is the tallest land animal and the largest ruminant, which is an animal that has multiple stomachs, and chews its cud like a cow. It stands about 18 feet tall and weighs about 4,000 pounds. The Rhinoceros is about 15 feet long and about 6 feet high at the shoulder and weighs about 5,000 pounds. Even with all that bulk, it can still outrun a human. The adult Hippopotamus is just about the same size and weight as the rhino, and it can also outrun a human. It is unpredictable and temperamental. Because of this, it has killed more people in Africa than any other animal. There are a couple of other large animals I'll mention here, the large, wild cattle, such as the Cape Buffalo in Africa, and the large Gaur cattle in India. All of these large animals are referred to by scientists as Mega-fauna, meaning "Large Animals" in Latin. We humans are small potatoes compared to them.

Now let's look at animals about our size. A good place to start would be domesticated farm animals. Some are bigger than us, and some are smaller. Horses, bulls and cows are bigger than us. Goats, sheep and pigs, are about our size. Ducks, geese, rabbits and chickens are smaller than us. Let's look at wild animals our size. Chimpanzees are about our size, smaller wild deer, large dogs, and wild pigs are too. Most birds are smaller than us, but one bird is larger, the Ostrich. A full grown male Ostrich is 9 feet tall and weighs about 300 pounds. There are other large flightless birds like the emu and the cassowary, but they weigh less than humans. But before we get any smaller, let's look at the world's largest insects. All insects are much smaller than humans but these giant insects are enough to fuel your nightmares for a month of sleepless nights. The Giant Huntsman Spider has a leg span of 12 inches, but it's a fairly slender spider. However, the Goliath Bird-eater tarantula has an 11-inch leg span, a body length of 4.7 inches and weighs over 6.2 ounces! Yikes!! That's almost a 1/3rd of a pound! The Goliath Beetle is the world's largest beetle at 4.5 inches long, and weighs 3.5 ounces. This one's not so bad. The world's largest moths are the Atlas Moth and the Queen Alexandra's Moth. Both of them have a wingspan of about 11 inches. I'll spare you the details of the world's largest Cockroach.

Instead, I'll tell you about the world's *smallest* insect. The world's smallest insect is the Fairy Fly, which is a very small parasitic wasp. The female is much larger than the male, but it is still only 1/2 to 1 millimeter long. The male is the world's smallest insect. His body is only 1/10 of a millimeter long. It would take 4 of them end to end to span the width of the period at the end of this sentence. That's about the size of a one-celled animal called a paramecium. This tiny male wasp has no eyes, no wings, and its legs are modified into suction cups to hold on to the female during mating. These little male wasps attach themselves to the female wasps as the females are being born from their eggs, and as soon as they impregnate the newborn female wasp, they lose all interest in them, and are useless after that, (sound familiar ladies?) Then when the female can fly, she looks for the eggs of other insects that have already been laid. Then, she punctures the eggs and lays her eggs inside the other insect's eggs. Then, the Fairy Fly's babies eat the other insect's eggs from the inside and hatch out of the other insect's eggs. The female Fairy Fly only lives a few days, but that's all she needs to lay her eggs. And you thought that your kids were too eager to grow up!

Most people never see one of these insects because they're almost too small to see. This is another good example of a Parallel Dimension. They're right beside us, but we don't even know they are there, and I'm pretty sure they don't notice us either, because they only have a few days to do so! There are many other creatures that are much smaller than humans, but how small do they get? Well, let's start with the largest, smallest animal from the whale family. The Vaquita, or "little cow" in Spanish, is the world's smallest porpoise. It's only 4 feet 5 inches long, smaller than an adult human. Its home range is limited to the north end of the Gulf of California. The smallest bird is the Bee Hummingbird of Cuba. It is 2 inches long and only weighs 1.8 grams. It's also the world's smallest warm-blooded vertebrate, weighing less than a U.S. Penny. The smallest mammal is a toss up between the Etruscan Shrew, which is 2 inches long and weighs 1.8 grams, and the Bumblebee Bat, which is 1.3 inches long and weighs 2 grams. The smallest rodent is the Baluchistan Pigmy Jerboa, which is basically a very small version of a Kangaroo Rat. Its body is only 1.7 inches long, with another 2 inches for a tail. Madame Berthe's Mouse Lemur is the world's smallest primate. It's only 3.6 inches long.

The Barbados thread Snake is the world's smallest snake, at only 3 to 4 inches long. It's said to be the same width as a piece of spaghetti noodle. The smallest fish in the world probably has the longest name of any fish in the world. Paedocyepris Progenetica is only about 1/3 of an inch long. It lives in Indonesia and is a member of the carp family.

The smallest *flowering plant* is Wolffia Arrhiza, or common Duckweed. Each plant is only 1 millimeter in diameter, and floats on the surface of quiet lakes or ponds. People who only see it from a distance usually call it pond scum, but it's really thousands of separate little round plants that are a favorite food for ducks. Every so often, these tiny plants sprout 3 tiny flowers for reproduction. The male Fairy Fly I mentioned above is probably the smallest multi-cellular animal that exists. If it's not, it's damned close! So, we'll use it to transition to the smallest microscopic animals, the largest first, then the smallest last.

The chart of microbe sizes I'm using next, is one I found on the internet. As a starting point, it uses a period at the end of a sentence, like this "." to compare to the sizes of different microbes. This "period", for you nitpickers out there, is a "Helvetica type, 12 - point size". This period is one half of a millimeter in diameter. The first microbe we'll look at is an Amoeba. The bodies of Amoebas constantly change shape, but in general, they are about 3/10ths of a millimeter wide. That's about half the size of our starting point *period*.

The next smallest one celled animal is the Paramecium. This is the critter that the male Fairy Fly is about the same size as. The Paramecium's shape doesn't change like an Amoeba's does, and it is shaped like the sole of a human's shoe. It's about 1/4 the diameter of our period. There are 15 microbes on this list, so I won't mention them all, just the interesting ones pertaining to size.

As we get smaller we come to the first and only fungus on the list, Bakers Yeast. Bakers Yeast is the fungus that makes bread rise. These little single celled fungi are only 1/100ths of a millimeter wide. The next smallest creatures on this size list are the Bacteria. The first and largest is Escherichia coli, more commonly known as E. coli. This bacterium lives in our guts and without it we couldn't digest our food. However, one form of E. coli causes serious food poisoning. Another group of bacterium, Lactobacillus, are also found in our guts to digest our food, and we also use it to make yogurt. E. coli, Lactobacillus, and other Bacteria are only about 2/1,000ths of a millimeter in diameter. It's an amazing fact that Bacteria are everywhere. They've even found them 11 miles deep in the Marianna's Trench at the bottom of the western Pacific Ocean. There are more bacteria on the earth, by weight, than all plants and animals combined.

Now let's get really small into the world of viruses. The first, and largest virus we'll look at is the smallpox virus. This virus is only 3/10,000ths of a millimeter wide. It's amazing that something so small could cause so much misery. There are 7 viruses on this list including the Rabies virus, the influenza virus, the Polio virus, and the smallest of all, the Rhinovirus, which causes the common cold. The Rhinovirus is an astounding 3/100,000ths of a millimeter wide. This ultra-tiny size is impossible for most of us to even imagine! I know I can't! Hmmmmmmm??

Let's see if I can explain these sizes in a more obvious way. Let's take 3 of these small critters from this list and compare them to each other. We'll take the largest, and first on the list, the Amoeba, then from about the middle of the list, the first Bacteria, E. coli, and then the last, and smallest on the list, the Rhinovirus. Oh, I almost forgot, let's include the period, because we can see it.

Period ..... = 0.5mm = largest
Amoeba... = 0.3mm
E. coli...... = 0.002mm
Rhinovirus = 0.00003mm = smallest

Now let's move all the dots to the right, 5 places, so that we'll be dealing with whole numbers for each size.

Period...... = 050,000mm = 164.041 feet
Amoeba.... = 030,000mm = 98.425 feet
E. coli...... = 000,200mm =   7.874 inches
Rhinovirus = 000,003mm =   0.118 inches

Let's look at the results. We enlarged the Rhinovirus size up to something we could see, 3 millimeters. At the same time, we enlarged all the other numbers the same amount. What an incredible difference in size. Now let's go over to the nearest mall parking lot and draw a circle 164.041 feet in diameter, and paint it white. That's the size of the period at the end of this sentence. Then let's draw another circle, inside the first one, that's 98.425 feet in diameter. We'll paint it red. That's the size of the Amoeba one-celled animal. Now let's draw another circle inside the second, red circle. Make it 7.874 inches this time. That's a little larger than the lid on a gallon can of paint. Paint this circle yellow. This is the size of the E. coli bacteria. Now, finally, draw a circle 0.118 inches or 1/10th of an inch in diameter in the yellow circle, this is less than half the diameter of a pencil eraser, and paint it black. This last, black circle is the Rhinovirus.

If this isn't a great example of Parallel Dimensions, I don't know what is. Isn't it amazing how something as small as a virus can cause so much trouble for beings so much larger than itself. Both creatures are completely unaware of each other's existence as living beings, except the larger is miserable but doesn't know why, and the smaller is overjoyed at it's good luck, and has no idea of the misery it's causing. All they know is that they're hungry. When this little 1/10th of an inch Rhinovirus circle bumps into a 98 foot Amoeba circle full of food, it just goes crazy, eating and reproducing. All it knows for sure is that it has hit the jackpot!

Scientists are not even sure that viruses are alive in the normal sense. Unlike other creatures, viruses can't reproduce among themselves like other creatures can. Other, larger creatures carry enough DNA in their bodies to mate with another of their own kind, who also has enough DNA in its body, so that the 2 creatures can mix their DNA to create a new individual. Viruses do have DNA in their bodies, but not enough to reproduce, so they have to invade other, larger creatures, and steal their DNA, cut out the parts the virus needs, and then it can reproduce. Some Amoeba, on the other hand, also cause some diseases, and all they know is that they are hungry, and they have found a big pile of food, so they start eating.

    The last thing I'd like to discuss is the different mind time speeds between large and small animals. As we all know, large animals move slower than small animals. Very small animals move very much faster than very large animals. A chipmunk compared to an elephant for instance. To us humans, who are mid-way in size between the two, the elephant looks like it moves in slow motion, while the chipmunk looks, to us, like it's moving much faster than normal. Smaller animals also live much shorter lives than larger animals. Although there doesn't seem to be any exact relationship between size and lifespan, in general, it appears to be true. Most insects only live from a few days, to a few weeks. Most mice only live a year or two. The short tailed shrew, which is sort of a miniature mouse, lives about 2 1/2 years. The Grey Squirrel lives about 14 years, and the American Beaver lives about 19 years. Large cats like the Tiger and the Leopard live about 20 years, and the Grizzly Bear lives about 47 years. The Asian Elephant lives about 78 years, the Killer Whale lives 40+ years, the Blue Whale lives about 110 years, and the Bowhead Whale lives 200+ years. Please be aware that all these ages and numbers can change at any time when the next study is published.

One last example of beings living in Parallel Dimensions are kids and adults. Anybody who has ever been a kid, or raised one as an adult, knows what I'm talking about. Even though they are both aware of each other, and they both live in the same place, they exist in separate, Parallel Dimensions. The Charlie Brown TV specials are a good example of this where the adult's speech is merely an occasional, disembodied, muffled trumpet blast that is promptly ignored.

Adults' and kids' brains have different time speeds. Kids run around in circles all day, and do more in one day than adults do in a week. I remember when I was 9 years old, my Mom and Dad and I moved into a new house in 1955. We lived there for 6 years, until I was 15. I loved living there and hated to move. It seemed like we had lived there a long, long time. When I was in my mid twenties, I began to realize how short a time we had lived there. When I looked back, it seemed as though we had lived there for 10 or 15 years, but it was only 6 years. I was amazed!

Now if it's obvious that brain speed is so different between adults and children of the same species, then how enormous must the brain speed be between say, a humming bird and an elephant? If an elephant feels something on the bottom of his hind foot, the sensation has to travel 20 feet to get to his brain, while if the same thing happened to a humming bird, the same sensation would only have to travel less than an inch. Taking into consideration the enormous body the elephant's nervous system has to manage, we can see why there would be an enormous difference in brain speed between these two animals, or any other two animals of tremendously different size. From here, I could keep going smaller to molecules, then down to individual atoms and then down to sub atomic particles, or I could go up to the largest living things, and go up from there to the sizes of the planets, and then up to the sizes of the stars and the galaxies, but I think I would be beating a dead horse.

I think I've made my point that the very huge and the very tiny, different dimensions, have always existed together, side by side, and we humans, have only been aware of the enormity of the situation for a few hundred years now, after the invention of the telescope and the microscope. Before that, we were very unaware of the very large and the very small in the heavens, and in and around ourselves. I could be wrong, of course, but I'm guessing that what I've discussed above are the only true Parallel Dimensions.

SPACE - TIME

As I mentioned in the previous article, Albert Einstein combined the 3 dimensions of space with 1 more dimension of time. He called this odd union *Space-Time*. People thought this was an odd union because the 3 dimensions of *Space* and the 1 dimension of *Time* didn't seem to have anything in common at all! But, it just seems that way. The one thing they do have in common is that neither of them exists.

Now let's look at space. Space is really just total nothingness. However, space is also filled with all the matter in the universe. This matter is spread all through the universe, but not evenly. It's concentrated mostly in separate clumps called galaxies, stars, planets and gas clouds with vast, comparatively empty, spaces between.

We can't measure empty space because there's nothing there to measure. Remember that when we measure something, we are comparing a physical thing of known size to a physical thing of unknown size. We compare a ruler of known physical length, to a physical board of unknown length. However, the clumps of matter inhabiting space are measurable. We can measure the width and the depth of the clumps, we can measure the number of the separate clumps, and we can measure the distances between the clumps, but we can't measure empty space if there is no matter, or something, in it to measure.

We also can't measure Time because there's no Time to measure, except in the minds of living beings. So, it sounds like Einstein made a big mistake, but he didn't. The remarkable thing about it was that he was right even though he didn't know exactly what Time was. If we substitute *Matter* for *Space-Time* we see that he was exactly right. Because *Matter* and its constantly reliable *Change of Physical Position* are real, and can be measured, but space, and time external to our bodies, are not real and cannot be measured. For this reason I think we should replace *Space -Time*, with *Matter-Copp*.

CHANGE AND MOTION

What is change and motion? The answer is that change and motion are identical. Then why are there 2 words to describe the same thing? The answer is human bias. Like any living creature, we are biased toward our own needs, abilities, and viewpoints. A bird flying the length of our backyard fence, forty feet away, is instantly visible to us as *Motion*. However, slowly sliding along the top of the same fence at the same forty feet away is a slug whose motion is invisible to us. When we look at the fence 2 hours later we see the same slug's position on the fence has *Changed* by about 4 feet, but we still can't see any *Motion*. Now if we walked out to the fence and got our face about 3 inches away from the slug, we could finally see that he was very *slowly in Motion*.

Another good example of Change and Motion is the clock. We can easily see the motion of the second hand, but the minute and the hour hands show no motion at all. But, like the previous example of the slug, if we get very close to the end of the minutes hand, we can see it very slowly turning. Try that with the hour hand and you won't see any motion at all. But if you go to the store and then run some errands, when you get back you'll notice that the hour hand has changed position. Now let's suppose that our clock had a fourth hand that counted 100ths of a second. Every time the second hand counted off 1 second this fourth hand would rotate 100 times. It would be a constant blur, like an airplane propeller. It would be too fast for us to notice any motion.

The point of all this is that everything in the universe is in constant motion in relation to everything else in the universe, even if we can't detect that motion by *seeing it*. A large boulder that has been lying on flat ground, not moving for tens of thousands of years, is in constant motion on the atomic level. That is what I mean by "Change of Physical Position".

So, we can say that *Motion is Change we can see, and Change is Motion that we can't see*! Simply Motion of different speeds!

TIME AND CHANGE

The Greek philosopher Heraclites lived about 500 years before Christ. He had such a pessimistic outlook about the universe and even life itself that people called him "The Weeping Philosopher".
He lamented that nothing seemed to be permanent. He even thought the Sun was created a-new every morning. His famous quote states "The only Constant in the universe is Change". If he had known about modern atomic theory and the expansion of the Universe, he could have claimed a major league "I Told You So"! There is an old

country song that I like from the 1930's called "Time Changes Everything". This assumption that Time is a basic property of the universe that *changes everything* seems to be pretty much universal. However, I think I can demonstrate that just the opposite is true. Change occurs on its own and Time has come into existence in living beings to help them cope with the constant amount of change, constantly occurring in and around them.

A parallel belief that was the opposite of what was true was what the ancient Greeks believed about sight. Hero, a mechanical genius, was a Greek engineer who wrote extensively on mechanical things. He was born about 20 years after the birth of Christ. He was the man who first demonstrated the power of steam, by turning boiling water into steam in an enclosed sphere with two bent tubes open to the outside air. When suspended over a fire by a chain, the steam leaving the sphere began to turn the sphere rapidly. Hero believed that eyesight was light-rays coming from the eyes and traveling at infinite velocity. He thought that when you got up in the morning your eyes were rested and therefore your eyesight was strong. As the day wore on your eyesight got more and more tired and finally when your eyesight was exhausted, it got dark. He obviously never went into a cave or a dungeon at high noon! Hero is a good example of how a very smart man can be very wrong. *External Light* causes *Eyesight* to work, and *External Change* causes *Time* to work. Just the opposite of what we once thought.

Change occurs all by itself and Time has come into existence, internally, in living beings, to help them cope with the constant amount of change occurring around them. In this chapter I will try to explain what Change really is, and its true relationship to Time. To do this, we need to identify something at a basic level that is common to all types of change throughout the universe. At first glance, this seems to be an impossible task when you consider all of the enormous changes that take place daily, just on the earth, much less in the entire universe. To understand this better, we need to define precisely what "change" is. This definition of change must describe all change in the universe that is possible to occur during the existence of the universe. When we think of all the change that occurs to us in a lifetime, and multiply that by billions of people and all animals, planets, stars, galaxies and just plain old debris floating around in space, it seems an impossible task. However, it just seems that way.

If we can discover a common denominator among all the change, we could make some headway. In other words what do all changes have in common with all other changes? I think a good way to think about this would be to describe a number of wildly different changes that seem to have absolutely nothing in common with each other. Then, try to find a common denominator among them all. Neither of us has all day here, so I'll limit the number of change examples to 10.

Change # 1. A man changes his mind.

Change # 2. The top and side of Mt. Saint Helens blows the mountain apart killing many people, thousands of animals, and destroys an entire forest.

Change # 3. An F-18 Hornet fighter jet loaded with bombs is catapulted off the deck of an aircraft carrier in the Indian Ocean.

Change # 4. Clouds appear in the morning sky and begin to grow.

Change # 5. The Sun goes down.

Change # 6. A seventeen-year-old girl puts on lipstick.

Change # 7. A supernova erupts in the Andromeda galaxy two million light years from earth that is brighter than the entire galaxy.

Change # 8. A mosquito lands on the bald - head of a man sleeping in a chair on his front porch, and begins to sip his blood.

Change # 9. A steel anvil sets in an ancient barn. It hasn't moved or been used in over a hundred years.

Change # 10. Someone buys a soda at a convenience store.

Now, I think these "changes" are varied enough to work with.

What could all these changes possibly have in common with each other? I believe the common denominator that binds all these changes together, is Motion, on the atomic and sub-atomic scale. In most of the above examples, there is a variety of physical changes that take place. Change # 7, a supernova, is a very large and powerful explosion that rips entire stars and their planets to shreds. Change #9, a stationary steel anvil that hasn't moved in a hundred years, seems to have absolutely nothing in common with a supernova, until we look at them on an atomic scale. Everything is always in motion on the atomic scale. The atoms in solid objects, like a boulder or a steel anvil, as in Change # 9, are a good example. Boulders and steel anvils seem very solid and unmovable to us. But on the atomic size level their atoms are constantly in motion, quivering in place, attached to each other by their electrons. If heated until liquid, these same atoms move around even faster, bumping into each other in a totally random, motion. If heated to a gas, the atoms really fly around with great energy and spread out in every direction unless captured in a container.

Robert Brown first detected this motion of the atoms in the early 1800's. He was a Scottish botanist who first discovered the nucleus in living cells. He also noticed that some unknown force was constantly jiggling around pollen grains in water, under high magnification. At first, being a Botanist, he thought that some mysterious *life force* in the pollen was causing the pollen grains to jiggle around. But when he looked at non-living dye particles he saw the dye particles being bumped around in the same way by the random motion of the water molecules. He had no idea what was causing this motion, but he did report it. In his honor, this motion is today called "Brownian Motion". It was the first observed physical evidence for the existence of atoms.

Since everything physical is made up of billions of these atoms, clumped together in various ways, I think that this atomic motion would make an excellent common denominator of change.

Now let's dissect just one atom. The simplest atom is the hydrogen atom so we'll start with it. The best way to see this atom is to make a model of it. Since atoms are so incredibly small, we'll have to make an incredibly large model just to see it. The nucleus of the hydrogen atom contains two particles, the neutron and the proton. They rotate closely around each other, and together, in our model, they are about the size of a standard baseball. That doesn't seem very large, but to be a complete atom, it has to have at least one electron in orbit around it.

In our model the electron is about the size of a grain of sand rotating around the baseball at a distance from our baseball nucleus of about two and one half miles. This would make the diameter of our model hydrogen atom 5 miles wide. It would almost be completely empty space inside the atom. The electron rotates around the nucleus in a random order, like when you roll up a ball of string. It rotates around the nucleus billions of times per second in our model. The electron would seem to be everywhere at once, like the spinning blades of an airplane propeller. Scientists call this an electron shell. Atoms bond to each other using these shells. What causes these electrons to constantly and reliably orbit the atoms so fast is completely unknown.

Brownian motion and the motion of electrons, is what we'll use as our common denominator of change.

Since these particles are constantly changing position in relation to each other, I'll call this common denominator *"Change Of Physical Position"*, or *"COPP"*. It works on a larger scale too. Two boys playing catch with a baseball, or the moon swinging around the earth, or two galaxies rotating around each other, are other examples of *"Change Of Physical Position"*.

Our next question is how does Time relate to this change of physical position? Well, the 20th century idea of Time, the "universal property of the universe" Time, is an attempt to explain exactly the same thing as the "change of physical position" explanation does.

So, we can say that in any process, formula, sentence, physical, or mathematical construction, the TIME explanation = the COPP explanation. Since we know that physical things exist, and that physical things constantly change position in relation to each other, we know that COPP is a real, physical thing. But is Time real?

I don't believe the physical, 20th century, *Universal Property of the Universe* type of Time is real at all, and I'll tell you why. In any chemical formula, or process, time is not used at all. Oxygen for instance, combines with other elements rapidly, as in a steel wool fire, or slowly, as in rusting iron. Both of these chemical reactions can be written down without using *Time* in the formula. The steel wool burns rapidly only because it's in the form of fine slivers of steel. Most of the volume of this steel is surrounded by oxygen ignited by the intense heat of a match, whereas a steel bar rusts slowly because most of the steel is not exposed to oxygen, moisture, and very little heat. But the process is the same. The same formula tells in exact detail the *Change Of Physical Position* that takes place among the various atoms in the chemical reaction without regard to time.

Another interesting example about Time and Change is a true story my Dad told me about. Dad worked as a maintenance man at Rollins College in Winter Park Florida during the 1960's. Rollins is located on the north shore of lake Virginia, in Winter Park. Back in the 1890's, they built some docks out into the lake, and by the 1960's, the old docks had mostly rotted away and they needed to be removed. Dad and some of the other maintenance men had to pull the old posts out of the lake. The old posts were misshapen to a soft, black, slimy point on top, but as they were pulled up out of the sand lake bottom they became firmer, and harder, until the bottom of the 80 year old posts looked like new. The bottoms were as hard and square as a new post and they still had the crayon marks that were put on them at the lumberyard 80 years ago. Dad was amazed by this huge difference.

Of course, the reason was that the bottom of the post was protected from oxygen and the other elements for 80 years or so and the top was exposed to the elements the same amount of Time. So, what did Time have to do with these changes? Well...nothing, it was just a matter of amounts. The amount of *Change Of Physical Position* acting on the top of the post was far greater than the amount that acted on the bottom of the post. It wasn't a matter of Time at all. It was a matter of amounts. If we have two explanations that equal the same thing, and one can't be shown to exist and the other explanation can be shown to exist, then the one that doesn't exist should be discarded. I think the old 20th Century *Universal Property of the Universe* type of Time should be discarded.

WHAT IS A CLOCK?

Another reason to discard "20th Century Time" is our old friend the clock. Scientists say that we don't know what a clock is because we don't know what it measures. We know it measures Time, but what is Time? How can we write an accurate description of a clock if we don't even know what it measures? Well, we can't. Or at least we couldn't until this book came along. Since

20th Century Time = COPP and COPP is real, and "20th Century Time" isn't real, then we can answer the question "what is a clock?" A clock is a machine that measures *"Change Of Physical Position"*, a *"COPP"* meter, or a *"Coppometer"*. A clock doesn't measure Time it measures *Change of Physical Position*. A clock is really just a mechanical version of the "Biological Time" that exists in our heads. The biological clock in our heads also measures *Change of Physical Position*.

    A ruler is an object of standard length that we compare to objects of an unknown length, so we can find out how long the unknown object is. The same is true of a clock. A clock is a machine that produces a very regular *Change Of Physical Position* to compare with other unknown amounts of *Change Of Physical Position* so we can find out how much *Change Of Physical Position* has occurred. We used our original, external clocks, the sun, moon and stars, the same way. We would compare the *Change Of Physical Position* of the sun from sunup on one day, till sunup the next day to the *Change Of Physical Position* that we accomplished here on earth. How many miles did we walk, how many bales of hay did we harvest, how many bricks did we make? In other words, we compare one known amount of change to another unknown amount of change. Just like a ruler compares one known amount of matter, a ruler, to an unknown amount of matter, a board of unknown length.

    Humans have a need to project their mental needs and desires onto the earth itself. Most animals also do this when they scent mark a territory for mating purposes or just living space. But humans, as usual, go to the extreme. We first built mechanical versions of our own body parts so our body parts wouldn't wear out so fast, but also because the mechanical body parts usually did a better job and lasted longer. A fist is still a good hammer for certain things, like adjusting electronic devices, but a steel hammer is more powerful, more versatile, and far more durable. A pair of cupped hands makes a good temporary water cup, but a cup can hold a lot of water with one hand while you use the other hand for something else. And the cup usually doesn't leak. A hand shaped like a rake works fine in soft soil, but if you're working all day in hard soil, you'll need a mechanical rake. A farmer seeing what part of the sky the sun is in is accurate enough for farming, but if he needs to be somewhere at 2:15 pm, he'll look at his watch.

When everyone was a farmer, or hunter, before there were cities, nobody needed to tell time very accurately. The sun, moon, stars and seasons told them all they needed to know. But when people began to gather in to towns and cities, things got complicated. When people began to inhabit cities, they didn't need to know the rhythm of the seasons as much anymore. Now they needed to coordinate their activities with all the other people they were shoulder to shoulder with all day. This meant that everybody needed to know time to the hour or minute. This is when clocks began to be built.

People didn't realize it then, but by building clocks they were trying to build a mechanical version of the biological clocks that already existed in their heads. We all know that people are slightly different from each other, and so are their biological clocks. By building a reliable mechanical clock, we can coordinate our activities in a much more practical way, especially when it comes to technological matters. While we're looking at clocks, let's look at man's first clock, the sun, moon and stars. Thousands of years ago, before cities or even farming had come along, we used the sun, moon and stars to keep track of daily, and seasonal life. The whole point of this is that we see reality through a narrow window. When we look at the full picture of change, we see that we look through a narrow window indeed! The full picture of change I'm talking about is all the change that occurs everywhere in the universe, from the monumentally large universal motions of the galaxies, to the monumentally small motions within the atoms. That's a lot of change going on, and we need to take it all into account to get the full picture of *Change Of Physical Position*.

THE ARROW OF TIME

Scientists trying to make sense of the mysteries of Time often mention "The Arrow of Time". They say the arrow of time is the fact that time only flows in one direction, from the past through the present and on into the future. Time never flows from the future into the past. Why is that? What causes Time to flow in the first place? Well, as I said in chapter 3, Time, our mental archivist, is constantly laying down memories in the near present. These memories are constantly pushed back into our mental past in an orderly manner, while our minds are constantly looking for expected future occurrences to arrive about every 30th of a second. This is the flow of time. There is no flow of Time in the non-living universe because there is no Time in the non-living universe.

A comedian, famous during the last decades of the 1900's, used to occasionally wear a gag arrow with a section cut out of the middle, where a "U" shaped wire connected the 2 halves of the arrow. He wore this fake arrow with the U shape upside down over his head so that it looked like he had an arrow shot through his skull from one ear to the other, while he acted as though nothing was there. Now, if he had twisted the arrow around so that the arrow point looked like it was coming out between his eyes, and the tail was entering the back of his skull, then you would have a perfect physical representation of the arrow of time.

Although, according to *The Biological Theory of Time*, time is confined to the inside of our bodies, our internal clock is programmed to think like the arrow. We think of the future as being ahead, and the past as being behind us. Our bodies are even built to reflect this attitude. Our eyes are on the front of our heads so we can see where we're going in the future. Our legs and bodies are designed to run in the direction our eyes are pointed. We can even see this attitude reflected in our speech. "Forge ahead, put the past behind you", "Full speed ahead", "Don't look back", "Hindsight is 20/20", "Back in the good old days", "Look forward to tomorrow", Stop living in the past!

Scientists of today sometimes ask why this *Arrow of Time* only goes in one direction. Why does it never go in reverse, so that a broken glass shattering as it hits the floor would rebound and re-assemble its broken pieces back into an unbroken glass again? According to *The Biological Theory of Time*, there is no *Arrow of Time* in the non-living universe, only in the living universe, in the minds of living beings. We can easily imagine such a thing as broken glasses reassembling themselves because most people today have seen movies, shown in reverse, of it appearing to happen. But this is not the way the real non-living world works. If no one had ever seen a backward movie of a breaking glass, this question would probably have never been asked.

TIME TRAVEL

Time Travel is a subject that is constantly speculated about in every other science fiction book, movie, or TV series ever made it seems. We think we might build a machine to travel back in Time to see the dinosaurs, cavemen, or to visit the Roman Empire at its zenith, or to visit the U.S. Army during WWII, and sell them your cell phone for a million bucks (they'd probably just take it away from you anyway). So where did all this Time Travel talk get started in the first place? I think the idea probably started when Isaac Newton began to think of Time as a real, physical thing, a sort of rigid lattice that the universe was mounted to. Maybe a real, physical Time could also be travelled, like a brick road.

Well, there were many books in the 1800's and earlier about something like *Time Travel*. There was the story of "Rip Van Winkle", by Washington Irving, who fell asleep in the woods and woke up twenty years in the future. Ebenezer Scrooge in "A Christmas Carol", by Charles Dickens, traveled into the past and the future accompanied by the Ghost of Christmas Past, and the Ghost of Christmas Future. Then there was "A Connecticut Yankee in King Arthur's Court", by Mark Twain, where the hero Hank Morgan, gets hit over the head by a crowbar in the America of the 1880's, and goes back in time to the England of the 520's. He convinces everybody that he is a magician, and with his 1880's industrial knowledge, starts to industrialize old England. He builds bicycles for the Knights of the Round Table, and defends himself from hostile knights by killing them with a revolver, and other feats of Yankee ingenuity.

But the book that gave birth to the modern notion of *Time Travel* was "The Time Machine" by H.G. Wells, in 1895. This was the first book to mention a machine that could carry a man, and precisely travel into the past or future, and return to the present. A movie was made of "The Time Machine" in 1960.

Is Time travel possible? Let's think about it. We know that travel is possible. We humans travel all the time in lots of different ways. The most common way we travel is just by walking. We might drive to work, but when we get there we usually walk around all day. We don't usually think of this as travel, but it is. We usually think of going on vacation as travel, and it is. We use a lot of machines when we travel. We use roller-skates, skateboards, bicycles, motorbikes, cars, buses, trains, ships, airplanes and rockets. We travel by way of land, sea, air, and space. Now let's see if we can find a common denominator among all these known forms of travel. In other words, what are we doing among all these forms of travel when we travel. The answer is that we are *Changing our Physical Position in relation to the earth when we travel.* Since Time does not exist outside our bodies as a physical reality, then Time cannot be traveled outside our bodies.

Time travel involves no change of physical position, so technically it's not travel. We can't travel to the intersection of Oak and Main streets if we're already *at* the intersection of Oak and Main streets. I'm convinced that time only exists in our individual minds. So the only traveling in time we could do would have to be in our heads. Although the space in our heads is a lot smaller than the world, it's still space. Scientists who study the brain tell us that our memories are real things. There are cells in the brain that somehow store the information that causes our brains to form a memory. Millions of our memories are stored in certain areas of our brains. When we remember things from our past, we are actually traveling through past experiences that are stored in physical places in our brains. It's sort of like walking among the shelves in our own, personal library. The change of physical position is small, but it's real, so when we remember the past, we are actually traveling to the only past that has ever been real. We can also travel to the future this way, by examining our past experiences that re-occur and we have come to expect to happen again. We all travel into the only future that has ever existed when we plan ahead for some event. So, Time Travel does exist in our heads. We don't even need a machine, just our memories, so let's go time traveling right now!

IS TIME LINEAR OR CYCLIC ?

Time is thought to flow in a straight, unending line by some people, but other people have the idea that Time flows in an unending circle, and that Time repeats itself over and over again. So, what is the true nature of Time? Is it linear or cyclic? Since true Time only exists in our heads, I would say that Time is linear. Time takes in our experiences, one at a time, and stores them in their order of their occurrence. Of course, Time records all our cyclic experiences such as day, night, months, seasons, and years, but it stores these experiences in a linear way. Now when we consider what we always thought Time to be, the *Change Of Physical Position* that is a basic property of all *Matter*, we see that all matter changes constantly in a cyclic way. From atoms with their electrons in orbit about them, to suns with their planets in orbit about them, to galaxies with their stars in orbit about them, most matter exists in a cyclic way. Also, we humans who are made of the same matter as everything else, live in a world dominated by constant cycles. The previously mentioned cycles of day, month, season, and year, plus the cycle of birth, growth, childhood, adulthood, old age, and death constantly repeat themselves without end. It's no wonder many people believe that Time is cyclic in nature. But the truth is that *Matter*, and it's built in *Change Of Physical Position*, is what is cyclic in nature, not *Time*.

If we look at all the natural cycles, we can see that all these cycles are not identical. Every day is different from every other day, every year is different from every other year, and so on with every other cycle we know of. So, we can say that even though the universe is overwhelmed with cycles, it is ultimately a universe ruled by a linear *Change Of Physical Position*, and that *Time*, in our minds, is linear as well.

CHAPTER 6

THE HISTORY OF THE CALENDAR

According to the ancient Roman myth, twin brothers Romulus and Remus founded the City of Rome. They were the sons of Rhea Silvia, a vestal virgin, a sort of pagan nun, who was required to swear herself to a life of chastity. One day, she gave birth to her twin sons Romulus and Remus who, she claimed, had been fathered by the Roman God Mars (a likely story). When the local King, Amulius, caught wind of this, he didn't believe her either, and had the twins tossed into the Tiber River to drown. Romulus and Remus were nothing if not lucky! They were saved by a she-wolf who suckled them back to health, and were fed solid food by a woodpecker (another likely story). Later a shepherd and his wife found them and raised them to manhood as shepherds (finally! something we can believe!). As adults, they proved to be natural leaders and decided to found a new city. There was trouble between them because they wanted to found their city in 2 different places. Well, tempers flew and Romulus killed Remus! So much for brotherly love, *A La Cain and Able.*

So, Romulus became King, and founded the city alone and named it Rome, after himself. He then founded the Senate and the first Legions. Also, he started the original Roman Calendar. The original Roman Calendar started at the founding of the city of Rome, which was about 735 years before the birth of Christ. Instead of the abbreviations B.C. and A.D. they used the abbreviation A.U.C., which meant *Ab Urbe Condita* in Latin, or "since the founding of the city" in English. So, Christ was born about 735 A.U.C.. Romulus' calendar was very different from today's calendar because it only had 10 months. King Romulus liked the number 10 a lot. He reasoned that people could keep track of the months easier if there were only ten months, because they could easily count them on their fingers.

Romulus named the months himself. He showed some imagination naming the first 4 months, but then his imagination ran out, and he started naming the rest, The Fifth Month, The Sixth Month, etc, etc, in Latin. These were the original names of the months in the original Roman Calendar.

First.Month..........Martis...........Mars
Second.Month......Aprilis..........Spring
Third.Month....,,,..Maius............The.Goddess.Maia
Fourth.Month.......Junius...........The.Goddess.Juno
Fifth.Month..........Quintilis
Sixth.Month.........Sextilis
Seventh.Month....September
Eighth.Month......October
Ninth.Month.........November
Tenth.Month........December

The weeks were about 8 days long. This was a Lunar Calendar, based on the phases of the moon. Their months were about the same length as our months today, so they had a couple of months of time left over at the end of the year that had no names or numbers associated with them. I guess you could call it "*float time*." Roman society at that time was very agrarian, they were mostly farmers, and after harvest in the fall, there wasn't much to do but survive until spring, so they saw no need to mark time during the winter months. The powers that be back then were shameless at using this *float time* to lengthen, shorten or juggle the calendar to their political and financial advantages. There were no neat little squares with numbers from 1 to 30 or 31 to display the month. The Romans simply had a vertical list of days for each month. The first day of every month was called the *Kalends* of that month. It's where we get the word Calendar. Then a list of numbers, largest first, smallest last, countdown to day one, called the *Nones* of that month. This was sort of the first week of the month. Then another week of numbers, largest first, smallest last, count down to the *Ides* of that month, which was the 15th, or the middle of the month. I'm sure you remember "Beware the Ides of March!" from the play *Julius Caesar*, by William Shakespeare. Then about 16 days, or 2 weeks, were put on the list, largest number first, smallest number last, and the last day was the *Kalends*, or first day, of the next month.

  This is a simplified version of the old Roman calendar. In reality, due to priests and politicians adding and subtracting days for political advantage, their calendar was pretty unruly. Also, the Romans had a fear of even numbers, which caused even more tinkering with the days of the months. After King Romulus died, King Numa Pompilius, the second King of Rome, took over and added 2 more months, called Januarius, named after Janus, the Roman god of beginnings and transitions, and Februarius, named after the purification ritual of Februa, which took place on February 15th, during the full moon, to fill out the year. The odd thing to us today, is that they added these 2 new months at the *end* of the year, after December. The reason was that back then farmers had for ages begun the year on the first day of spring, or planting season, which was on March 1st.

The historical record is vague, but they probably added the 2 new months as more and more people began to live in cities. The farmers were basically "dead in the water" during the coldest winter months, but people in the cities did business all year long, and they needed to be able to mark time all year long. Later, Januarius and Februarius were moved to the beginning of the year. The reasons for this change are also vague. The only thing historians seem sure of is that it happened, and that today, January and February are the 1st and 2nd months of the year. The main problem with the calendar back then (aside from the politicians) was that nobody knew exactly how long the year was. This, and the politicians, caused the old Roman calendar to gradually get out of sync with the actual year and to be reformed over the centuries to give us the calendar the world uses today.

The first great reform of the old Roman calendar was called the Julian calendar, named after Julius Caesar, the first Emperor of Rome. He made changes to the calendar to bring it back in line with the seasons because, at that time, it was a few months (months?!) out of sync with the real year. Julius Caesar, while floating down the Nile with Queen Cleopatra, on her royal barge, once took a break from fooling around with Cleo, and struck up a conversation with one of the old wise men on board, Sosigenes of Alexandria. It was a big barge with lots of rooms. The old wise man told Caesar that the Egyptian calendar was several thousand years old and was very accurate because it was based on the movement of the Sun, and the flooding of the Nile. This inspired Julius Caesar to finally do something about the old Roman calendar. He called together the best and brightest of the Roman worlds philosophers and mathematicians to guide him in setting the Roman Calendar on the right path. When they were done, several changes were made. The Calendar would now be based on the motion of the sun, and the beginning of the year was moved from March 1st to January 1st, so the year would begin at the winter solstice. They finally decided that the year was 365 1/4 days long. To make up for the fraction of a day they made a leap year of 366 days every 4 years, which we still do today.

Old Roman	Julian	Todays
1. Januarius	Januarius	January
2. Februarius	Februarius	February
3. Martius	Martius	March
4. Aprilis	Aprilis	April
5. Maius	Maius	May
6. Junius	Junius	June
7. Quintilis	Julius	July
8. Sextilis	Augustus	August
9. September	September	September
10. October	October	October
11. November	November	November
12. December	December	December

You'll notice above that Quintilis and Sextilis were renamed July and August by the Roman Senate in honor of Julius Caesar and his son Augustus Caesar. I'm sure that Julius and Augustus had absolutely nothing to do with this decision! Hmmmmm? This renaming of the months and days got to be a fad amongst the ruling class of the time, and after it got out of hand, cooler heads prevailed to prevent any more than Quintilis and Sextilis to be renamed. One of the most unusual things they did was to add 3 extra months to the year 46 B.C. only. There was one extra month near February, and 2 extra months between November and December. This made 46 B.C. 455 days long. The Romans appropriately called this "The Year of Confusion". All this tinkering brought the Roman Calendar into sync with the true year, and made it one of the most accurate calendars in the world.

The Julian calendar was launched on January 1st, 45 B.C., the year after the Year of Confusion. After the Year of Confusion was over, the new Julian calendar was much improved, and much more accurate. However, it was still slightly out of sync with the true year. But this error was small enough that the Julian calendar was used for the next 1600 years.

The next great change to the Roman calendar came about 350 years after the death of Julius Caesar, when Constantine the Great became the first Christian Emperor of the Roman Empire. While praying for victory in an upcoming battle that would determine whether or not he would become Emperor, Constantine said that he saw a flaming cross in the sky over the sun with the words, in Greek, saying "In This Sign Conquer". That night he also dreamed he heard a voice tell him if he painted a symbol of Christ on the shields of his soldiers, he would win the battle, which he did. The Roman Empire, and Europe, would never be the same again. Constantine became Emperor of Rome in 312 A.D. after winning the battle of the *Milvian Bridge*. Constantine inherited a bureaucratic and economic mess in the Roman Empire that was held together by an all-powerful army. The empire needed a strong and effective leader, which Constantine turned out to be. He ruled for 31 years and made many changes to prop up and rejuvenate the faltering empire. Some of these changes were made to the Julian calendar.

In 321 A.D., nine years after becoming emperor, Constantine introduced Sunday as the Sabbath day, and the first day of a new, seven-day week. He also was the first to put Christian holidays on the calendar, such as Christmas and Easter. This 7-day week idea was quite popular in Rome at the time, having been invented by the Babylonians about 700 B.C.

The Babylonians also named each day of the week after their gods, which the Romans did also, substituting Roman Gods for the Babylonian ones. People back then had a lot of faith in astrology, and the number 7 was considered a mystical, almost sacred number. There were 7 planets in the sky that the Romans considered Gods, The Sun, The Moon, Mercury, Venus, Mars, Jupiter, and Saturn.

The Roman / English, names for the days and planets were:

Roman ----------------English

Sol .... Sun
Luna .... Moon
Mars .... Mars
Mercurius .... Mercury
Jupiter .... Jupiter
Venus ....Venus
Saturnus ....Saturn

The modern romance languages, such as Spanish, are very similar to the old Roman names for the days of the week.

Roman............Spanish

Sol ....Domingo
Luna ....Lunes
Mars ....Martes
Mercurius..........Miercoles
Jupiter ....Jueves
Venus ....Viernes
Saturnus ....Sabato

The Anglo-Saxons in England during the 400's A.D., who were basically just a bunch of transplanted Germans, also wanted to get in on this craze of naming the days of the week so they could appear more Roman. So, they mostly used their own Scandinavian/German Gods. Tiw was the Norse God of war, Odin, or Wodin, was the Chief of all the Norse Gods, Thor was the Norse God of thunder, and Freya, was Thor's *wife*. The English word, Day, comes from the Saxon word *To Burn*. So, the lists below are:

Roman............. Anglo-Saxon.........English.

Sol................. Sun-Day............ Sunday
Luna................Moon-Day............Monday
Mars ....Tiw's-Day ....Tuesday
Mercurius..........Wodin'sDay..........Wednesday
Jupiter.............Thor's-Day............Thursday
Venus...............Freya-Day...........Friday
Saturnus............Saturn-Day...........Saturday

In 325 A.D., thirteen years after becoming Emperor, Constantine convened the Council of Nicaea, which was the first major Christian council to unify the scattered sects of Christianity into the state religion of the Roman Empire. This became the Universal, or Catholic, Church, which far outlived the Roman Empire, and is still with us today. One of the main topics of the council was how to determine the date of Easter, which falls on a different day each year. This kept the debate about the accuracy of the Julian calendar alive for over a thousand years to come. In 525, an abbot named Dionysius, aged 26, was asked by Pope John to improve the method of determining the date of Easter. When Dionysius finished his tables he wrote the year as 531 A.D., which was an abbreviation he invented that stood for Anno Domini in Latin, or Year of Our Lord, in English. Dionysius was the first to ever do this, but it was his friend Cassiodorus who was the first to publish this A.D. notation in a textbook in 562 A.D. This A.D. notation took a while to catch on. The other Italians accepted it by the late 500's, the Britons adapted it in the 600's, Gaul, or France, started using it in the 700's, and outlying areas, like Spain, finally accepted it in the 1300's. It wasn't until 1627 that a French astronomer, named Denis Petau, invented the opposite notation of B.C., or Before Christ.

This B.C.—A.D. business was especially important when the calendar became a Christian calendar and started counting up from year 1 into the future, and up from year one into the past, because it was the only way to tell the difference between the 2 sets of identical numbers. This would be especially important for historians. There was a 500 B.C. and a 500 A.D. for instance. In 825 A.D., an Arab scientist named Al-kwarizmi, known as The Sage of Baghdad, wrote a small booklet about mathematics that, when translated into Latin, introduced Europe to the numbers of India. These are the numbers we use today, 1 thru 9 and Zero. The Roman calendar up till then had used Roman numbers, which were much clumsier to use. These numbers from India were called Arabic numerals in Europe, because they learned of them through the Arabs.

The last great change to the calendar came along over a thousand years after the fall of Rome. Pope Gregory, in 1570 A.D., had become aware that the calendar was still not keeping pace with the actual seasons and needed further reform. He assembled a large group of mathematicians and scientific experts of his day, and put them to work trying to figure out a plan to fix the calendar. It took about 10 years of deliberation until they finally agreed on the last detail of the Gregorian calendar reform.

Julius Caesar and his councilors had decided that the year was 365 days and 6 hours long. The true length however is 365 days, 5 hours, 48 minutes and 45 seconds, about 10 minutes shorter than what Julius Caesar and his councilors thought. That's about 10 minutes a year short of the Julian calendar. In a thousand years that's about 7 days

short of the true year, and at the year 2000 it would be
off by 2 weeks! The Pope's problem was mostly one of accuracy. It was a problem of mathematics. What they finally did, was introduce a set of rules that inserted a number of leap years at certain places through the coming centuries that would keep the calendar very precisely in sync with the real year. Also, they needed to get rid of 10 extra days that had accumulated since Julius Caesar's time. They thought about getting rid of these extra days gradually, but instead, they decided to chop them out of the calendar all at once......Surprise! Finally, after 1600 years, in 1582, Pope Gregory's new, improved calendar went into effect.

October 4th, 1582, was the last day of the Julian Calendar, then, all of a sudden the next day was ten days later, October 15, 1582, the first day of the Gregorian Calendar. Everyone had lost ten days of their lives! People were really disturbed about this! Many were convinced that their lives were shortened by 10 days! But, after a few introductory hiccups, the new calendar began to take hold. The main reason was increased accuracy. The Julian calendar only took 128 years to go 1 day off the true year. The Gregorian calendar takes 3,226 years to go 1 day off the true year. This was a huge improvement in accuracy! Only the Vatican and a few Catholic states accepted the new calendar at first, but by 1584, most Catholic countries in Europe had accepted the new calendar. England and the American colonies accepted it in 1752, and Germany did by

1775. Because the British Empire spanned the globe at that time, their new calendar did too. Other European countries with overseas colonies took the new calendar with them also. All these world colonies were developing an enormous world trade among nations, which was what all the colonies were about anyway. Thus, the world had a new, accurate calendar to facilitate world trade and coordination. But it didn't happen overnight. Japan accepted the new calendar in 1873, Russia in 1917, and China not until 1949. So, today, the default world calendar for trade, business, and world coordination is the Gregorian calendar.

As convoluted as this chapter is concerning the development of the calendar, in reality, it was far more complex a process than I have presented here. I left out many important people and their roles in the reform of the calendar to avoid putting you, dear reader, to sleep. I just wanted to present a brief summary of the situation, to prepare you for the next chapter. In preparing for the next chapter, I wanted to show that the calendar has been completely man-made in a long twisting and turning manner throughout the centuries.

God invented Time.
Man invented the calendar!

CHAPTER 7

A SLIGHTLY NEW CALENDAR

Mankind has invented many calendars over the centuries. Some are based on the motions of the moon. Some are based on the motions of the sun. Many are based on certain religions, and some are more or less accurate than others. In the last chapter, I showed the development of the calendar over the last 2,000 years, and how accurate it has become. But in this chapter, we will concentrate on the most commonly used calendar in the world today, the modern version of the old Roman calendar, the Gregorian calendar. I believe we can improve our calendar with a few basic changes that would make it much easier to for everyone to use. I'm sure a lot of people would think I'm insane to suggest that we change the calendar. They would accuse me of trying to play God. But we must remember:

God invented Time.
Man invented the calendar!

I'm only a man trying to clean up a small mess made by other men. Time has served us very well over the ages. The calendar doesn't work as well as Time, but it could be better. The Gregorian calendar is still a bit of a mess. But it's a mess we're used to. To prove this, all I have to do is recite the old poem that was written to help us remember the number of days in each month.

"Thirty days hath September,
blah-blah-blah-blah, can't remember!"

I'll bet you can't remember it either. I've always wondered why the months have four weeks and a few days. And then to really confuse us they add a day to February every four years, and it still doesn't have as many days as the other months. Also, why do half the other months have 30 days, and the other half has 31? Then every month starts and ends on a different day! We don't have enough to think about all year? Do we really need this? I don't think so!

I believe we can improve the calendar a tremendous amount with a few simple changes. None of these changes will have anything to do with the mathematics that made the Gregorian calendar so accurate and successful. The changes won't change the length of the year at all. These changes also will not change the mathematical rules for the Gregorian calendar leap years, which have made the Gregorian calendar so accurate. However, there would be 2 rather big changes! The first big change would be to break the seven-day week cycle once every year. The second big change would be the addition of an extra month making the calendar a 13-month calendar.

When I first thought about this, I wondered why we couldn't have exactly 4 full weeks in every month instead of 4 weeks and several odd days at the end of each month? So, I took my calculator and divided 365 days by 28 days (exactly 4 weeks). I was amazed to find that there would be 13 months of exactly 28 days each with only 1 day left over! Only 1 day? Amazing! This arrangement of the months and days would be the same every year forever, making it a perpetual calendar.

First let's look at breaking the seven-day week cycle. This seven-day cycle had flowed unbroken for over a thousand years since the seven day week was utilized by Julius Caesar in 46 B.C. when his Julian calendar first came into use, except for the 10 days removed in 1582. This, plus 12 months with their irregular number of days, causes the year to start and end on any day of the week, and is totally unpredictable by the average person. If we started the year on a Sunday, and ran this seven-day cycle for 365 days, the 365th day would also be a Sunday. If we then started all over again with another Sunday for the next year then that year would also end on a Sunday. Every year would end and begin with Sunday. The last Sunday of the last year, would be next to the first Sunday of the next year, making a double Sunday at the beginning and end of each year. This would occur every year until leap year, when the last day of the year would be an extra day, a Monday sandwiched between 2 Sundays. We could call it leap Monday. Or we could make it a leap Sunday. That way we would have a triple Sunday every leap year, and there would be no fraction of a week in the calendar. This would be a perpetual calendar.

90

Diagram # 1. The 28 Day Month

SUN	MON	TUE	WED	THU	FRI	SAT
1 [29]	2 [30]	3 [31]	4 [32]	5 [33]	6 [34]	7 [35]
8 [36]	9 [37]	10 [38]	11 [39]	12 [40]	13 [41]	14 [42]
15 [43]	16 [44]	17 [45]	18 [46]	19 [47]	20 [48]	21 [49]
22 [50]	23 [51]	24 [52]	25 [53]	26 [54]	27 [55]	28 [56]
29 SUN 365th Day	30 SUN Leap Day 366th Day	1 SUN New Years Day — This is the First Day of the Next Year				

The 29th and 30th are in December Only !

The above is the Perpetual Month in the 13 Month Calendar

The small number in the upper right corner of each date box is the day of the year

The example above is for February

This would occur every year until leap year, when the last day of the year would be an extra day, a Monday sandwiched between 2 Sundays. We could call it leap Monday. Or we could make it a leap Sunday. That way we would have a triple Sunday every leap year, and there would be no fraction of a week in the calendar. This would be a perpetual calendar.

This would be intolerable to many religious people who follow the Bible's teaching to work for 6 days and rest on the 7th day. But, 99% of the time during the year, we would still be observing the seven day week cycle. Also, religious people could still follow the original Gregorian calendar if they wanted. Many religions today follow calendars specific to their religious beliefs. It wouldn't be the first time this has happened. For example, the Seventh Day Adventist church celebrates the Sabbath on Saturday. This has been going on since the Roman Emperor Constantine declared in 321 A.D. that Sunday, instead of Saturday, was to be the Sabbath. There are also many other religious calendars in use in the world today. The Jewish calendar, which still observes the Sabbath on Saturday, the Islamic calendar, the Eastern Orthodox Christians still use the old Julian calendar, and there are many more. People who worry about this break in the week cycle are probably still not aware that our current calendar had 10 days cut out of it on October 5, 1582, when the present calendar went into effect.

That means that nobody who has followed today's calendar has worshiped on a Sunday, or even Saturday, in 500 years. This fact doesn't seem to bother most people today. Having said that, I still think people would grow to love the great regularity and predictability of a 13 month, 28 days a month calendar. And my final argument for breaking the 7-day week cycle, is that the first 2 men to propose this idea were men of the cloth. The Reverend Huge Jones of William and Mary College in Virginia was the first in 1745, and the second was Abbot Marco Mastrofini of Rome, Italy in 1834. Surely these 2 men were devout Christians, yet they saw no problem with breaking the seven-day cycle a little each year.

The second big change in the calendar would be the addition of a new month. The calendar would become a 13-month calendar. I haven't added another month of time, I've just collected all those few extra days lying around at the end of every month, beyond 28 days, and bundled them into an extra, 28-day month. The problem is where do we put it, and what do we call it? I think the trick to making such a big change to the calendar, as a new month, would be to make it as unobtrusive as possible, regarding what people are used to. If we put it at the beginning, the end, or in the middle, it would be too obvious a change. I think we should wedge it in between August and September. That's about the end of summer when people are sick of the hot weather and are looking forward to the cool weather of fall. Vacation time is ending, school is starting, and the holidays are a few months away. These distractions could easily cover up the extra distraction of that annoying new month. Also, if we look at the last 3 letters of each month we can see a bit of a pattern.

ARY	RCH	BER
ARY	RIL	BER
	MAY	BER
	UNE	BER
	UST	

As you can see, the first 2 months both end in ARY. The middle bunch of months all end differently, then the last 4 all end in BER. If we put the new month between August and September, and made it end in BER, there would be as many BER months as the different months. This way, the new month might fit in fairly unnoticed. But what do we call it? We could call it anything really, but remember, we're trying to be unobtrusive. It should be longer than May, the shortest month, and shorter than September, the longest month. August has the middle length with 6 letters. Also, it should sound something like the existing months if possible. How about Autember? It would be a blend of August and September, it sounds a bit like Autumn, which is about when it would occur, and Autember sounds pretty good too, almost like it belongs there.

This break in tradition would allow us to have a perpetual calendar forever. Not only that, but by being able to remember the numbers of the days of each week for 1 month, we would be able to remember all the months of the year forever! This would make it far easier for people to coordinate with each other personally, or in business. We could then stick that one extra day to the end of December, making December 29 days long. Then, on leap year, we could make the leap day December 30. By putting these extra days at the end of the year instead of in the middle of the year somewhere, like we do today with February, we could avoid a lot of confusion. People would be confused at that time of year anyway because of the mental jumble of the holidays, and the new year, so an extra day or two at the end of the year wouldn't be that much more confusing.

Every month would then be identical to every other month forever, except for the different names of each month, and the different holidays in each month. December would also be identical to all the other months, except for 1 or 2 days at the end of December.

Another important thing this change would do is make every January 1st a Sunday. Then the first day of every month and of every week would be a Sunday. The end of every week and every month would be a Saturday. Then December 29th, the last day of the year, would also be a Sunday. The first week of every month would begin on Sunday the 1st, and end on Saturday the 7th. The second week of every month would begin on Sunday the 8th, and end on Saturday the 14th. The third week of every month would begin on Sunday the 15th, and end on Saturday the 21st. The fourth week of every month would begin on Sunday the 22nd, and end on Saturday the 28th.

In my new calendar, every year would start on Sunday, January 1st, and end on Sunday, December 29th. Then, the next day would be Sunday, January 1st of the next year. There would be no week between these 2 Sundays, just 2 Sundays in a row, a Double Sunday. However, on leap year there would be December 30th, a Leap Sunday between these 2 normal Sundays. A triple Sunday! It's not perfect, of course, but it's awfully close. I think the advantages would far outweigh the disadvantages.

ADVANTAGES

The advantages of a 13-month calendar would be many. The calendar would be much more predictable by everybody due to its great regularity. If you can remember the days and weeks, and their numbers for one month, you could remember all the days and months of every year forever. Bookkeeping would be far easier for businesses, and even personal bookkeeping would be easier. The average person might not consider making things easier for businesses much of an advantage for them. They would be wrong.

If things are easier and more productive for businesses, then they are more able and willing to hire more people, and pay them a higher wage. If an easier to use calendar could be used around the world, everyone around the world would be more prosperous.

George Eastman was a self made man who founded the Eastman Kodak company. He was one of those people who gave to the world far more than the world could ever repay him for. He introduced photography for everyone throughout the world. He introduced profit sharing for his employees when it was unheard of. He gave away tens of millions of his own money to universities and other organizations. He was also a huge promoter of the 13- month calendar. From 1924 to 1989, he ran The Eastman Kodak Company using this 13-month calendar because it made the accounting in the company so much easier.

He felt that in the modern international world of business that the accounting of Time periods was a lot more difficult than it needed to be because of the variable Time periods in the calendar. The most annoying to him were the month, the quarter, and the half-year. Of these three, the month was the most annoying because they're always a different length of time.

For instance, the number of days in the months of the year 2017 are, – 31, 28, 31, 30, 31, 30, 31, 31, 30, 31, 30, 31.

In the example above there is only one month, February, that is exactly 4 weeks long. Another difficult problem for bookkeepers is that some months have 4 weeks and others have 5 weeks. All these varied Time problems make it extremely difficult for businesses to tell if one month is more or less profitable than another month. This also makes it more difficult to tell if a certain business practice is more or less profitable than another business practice.

Eastman thought that the key to solving this accounting problem would be to have a perpetual month of exactly 4 weeks. To do this an extra month of 4 weeks would need to be added to the calendar. The year would be 13 months long, but the year would still be the same number of days long. Eastman said that a 13-month calendar would give everyone a faster money turnover during the year. Businesses would send out 13 instead of 12 monthly bills. People who get paid monthly would get paid 13, instead of 12, times a year.

During the first part of the 1900's George Eastman and The League of Nations promoted the Moses Cotsworth Calendar as the best 13-month calendar that they thought the world should adopt.

To read more about this see below

A LETTER FROM GEORGE EASTMAN

Nation's Business, May, 1926, p.42-46

THE IMPORTANCE OF CALENDAR REFORMTO THE BUSINESS WORLD

The calendar makers wouldn't like this, but you could go without buying a new calendar every year. All you'd need would be a plaque on the wall showing the perpetual month. But since every month has different holidays, and every month has a different name, and every year has a different number, people would still buy calendars. I remember my mother used to write down everything on her calendar, the weather, when she got a letter, when bills were due, Doctors appointments, anything and everything. The calendar makers could even make those perpetual month calendar plaques I mentioned above.

 The people who take all those pictures of little fuzzy bunnies, kitties, puppies, and nude girls, would be able to sell one more picture to the calendar makers every year. Everyone would prosper. Many holidays would change for the better. George Eastman, who founded Kodak, says that a 13-month calendar would enable Easter to be a fixed holiday. Halloween would always fall on the 28th and last day of October, which would always be on a Saturday. No more weeknight Halloweens. I think everybody would enjoy that more. Thanksgiving would always fall on Thursday the 26th of November during the last week of the month like it always does. Christmas would always occur on the 25th of December, on the last Wednesday of the last week of the month. No more bouncing around the month for Christmas, and no more, early Christmas.

 If your boss gave you the week of Christmas off with pay, he would pay you for 5 days, but when you count the weekends and New years Day, you would actually get 10 days off or eleven days on leap year. If your boss makes you work the 2 days before Christmas, you'd still get 8 days off. If your boss makes you work the 2 days before and after Christmas, then, well, maybe you need a new boss, unless your boss is paying you $1500 an hour to work Christmas week. In that case, MERRY CHRISTMAS!!!

DISADVANTAGES

Just to keep you cynics happy, I guess I'll have to be fair and think up a few disadvantages. So here goes.

It would be a nuisance to convert to a new calendar! However, with today's technology, it would be easy to display 2 calendars side by side on your electronic hand gadget to show the current date. It could also be done easily on a paper calendar by printing the numbered day of the year, in red, in the upper right side of each little box showing the day of the month. You could look at the current date on the old calendar, and then find the current date on the new calendar by comparing the numbered days of the year in the 2 boxes. Like I said, it would be a nuisance. But look at it this way, it would be a nuisance to sell your old car that got 5 miles to the gallon, and then buy a new car that got 40 miles to the gallon, but you'd get used to the new car really fast.

All the dates would change! Well, all the dates after January 28th would change. Here's where the numbered day of the year would come in handy again. Just look up your Birthday, Anniversary, graduation day, or any day on the old calendar and then look for the same numbered day of the year on the new calendar. A lot of people would not like having to change their birthday. My birthday would be off by 13 days in the new calendar. Not because I'd lost 13 days, but because the new months were shorter, my birthday would still be on the same day of the year. Of course, you could just continue to celebrate it using the old date on the new calendar if you wanted. In the old calendar, around the middle of August you'd only be about 2 weeks out of sync with the new calendar. Then the new month of Autember would be inserted in the new calendar and the months would begin to come back into alignment again, until on December 31st (old) and December 29th (new) they both would end the year at the exact same time. Now, if you're an old geezer like me you might just want to forget your birthday altogether. Or maybe the opposite is true. Maybe you love celebrating your birthday and would do it twice a year if you could! Here's your chance!

    National holidays would change dates. The fourth of July would no longer be on Independence Day. The new date would be on Tuesday, July 17th, every year. It could be a 4-day weekend forever! Being a patriotic American, this bothers me. I've always thought of the Fourth of July as Independence Day. But it would, at least, be celebrated on the correct day of the year. The same thing would happen with Pearl Harbor Day. It would be celebrated on Dec 5th, not Dec 7th, but still on the correct day of the year.

    Now let's look at the silly stuff. Not only would there be 13 months in the year, there would be 13 Friday the 13th's in the year. And to cap it all off we'd get the double whammy of a Friday the 13th on the 13th month at the end of the year! This would happen every month FOREVER!! This would be a boon for the superstitious among us. They'd get to call in sick at work every Friday the 13th, Well.... at least for a while. The good thing about it would be that after 10 years or so of Friday the 13th's every month, it might occur to the superstitious that Friday the 13th being unlucky is a load of crap! Remember that the Gregorian calendar of today is more than just a Religious Calendar it's the world's calendar. This new calendar would be a huge improvement in peoples and nations' ability to coordinate with each other, in business and other activities around the world. Not only that, but by being able to remember the numbers of the days of each week for 1 month, we would be able to remember all the months of the year forever! This would make it far easier for people to coordinate with each other, personally and in business.

Near the end of writing this book I was taken completely by surprise by a new development in astronomy that would be another advantage to a 13-month calendar. NASA has determined that there are actually 13 constellations in the zodiac, not 12. Apparently, the Babylonians, who invented astrology, jammed 2 constellations together into the same month so that they could make the zodiac conform to a 12-month lunar year. NASA tells us that Ophiuchus (Oh-fee-OO-cuss), "The Serpent Bearer" is the new 13th constellation in the zodiac. For you non-stargazers, the zodiac is a line through the sky that all the planets follow in their orbits around the sun. The planets all follow the sun's equatorial plane in their orbits. In other words, they all orbit on the same plane, like marbles rolling in a circle around a flat tabletop. Because of this, they never vary very much from this line in the sky. For example, you never see a planet in the far north or south of the sky. The astrologers are all in a tizzy about this, claiming that Ophiuchus is not a real sign of the zodiac. Personally, I think having a sign of the zodiac named the serpent bearer makes perfect sense. Astrologers have been selling this "snake oil" to the public for thousands of years. I guess they think that the snake bearer cuts a little too close to home and might somehow "blow their cover."

Now let's go back to our little poem to remember the number of days in the month. Can you remember it? Here it is below.

> Thirty days hath September,
> April, June, and November.
> All the rest have 31,
> Except for February all alone.
> It has 28 each year,
> but 29 each leap year.

I'm sure you could remember the above by heart if you spent a lot of time working at it, but who needs the aggravation? So, if my slight calendar change were adopted, it would be much easier to remember using my new poem below.

> Twenty eight days hath September,
> So do the rest, except December.
> December has 29, my dear,
> Except for 30, on leap year.

How do you like my new little calendar poem? I love it! It has a really nice rhythm to it. But, the most wonderful, beautiful thing about my new poem is that it's totally un-necessary! Who, in their right mind, can't remember one extra day a year, placed at the very end of a year with 13 identical months? If you can't remember a little detail like that, you deserve to have to memorize a poem! So, that's my plan to slightly, or enormously, depending on your opinion, improve the calendar.

CHAPTER 8

A NEW TIMELINE FOR HISTORY

I believe a new timeline for history would also be a good idea. As it is now, the years get smaller when we go back into the past, which is easy to remember. But then, we come to the birth of Christ, and all of a sudden everything gets backwards, and the years get bigger again. I'm sure the historians are used to this, but to the rest of us, it's just a pain in the butt! Another couple of irritating things about the current timeline is that first, none of the 12 apostles ever bothered to ask Jesus when he was born! Nobody today even knows what *YEAR* he was born in! Historians think he was born somewhere between 2 and 4 B.C. When the Apostles were asked this question, they said that they didn't ask Jesus when he was born. They didn't care about time! It wasn't important they said. Sounds just like a bunch of guys trying to cover their rear ends for not asking the obvious question. Besides, they probably thought Jesus would be right back in a few weeks or months anyway, so no need to think about time. I've often thought that if the 12 Apostles had been women they would have hounded Jesus, or Mary, or both, for Months until they found out the year, day, hour and minute of his birth. Then, every year, they would have sent him a birthday card, signed by all of them, and baked him a Birthday Cake, and today we would know exactly when to celebrate his birthday. But, they weren't women, they didn't ask, and we don't know the date, or even the year, when Christ was born.

There are 3 things I'd change about the current timeline. First I would add a *Zero Year* to the timeline. This would not change any of our current dates of the last 2000 years because this one extra year would go backwards in time from the year one. Then I would add a *Zero Century* to the timeline also. This *Zero Year* and *Zero Century* would start at exactly the same time. When the church changed the calendar to start from the time Christ was born, it had never heard of *Zero*, so they just started the timeline on the year 1. We still have problems with this today. When the calendar turned 2000 it wasn't 2000 years since the birth of Christ. It was 1999 years since the birth of Christ. We had to wait an entire year until 2001 to mark the second millennium because our timeline did not start from a zero year. This is really confusing!

We can visualize the problem by looking at a horizontal row of empty squares, like on the calendar, only empty. Let's draw 6 of them. Each one represents a year or a century. Now over each square starting from the left put a number 1 over the top middle of the first square on the left. Then a 2 over the middle of the next square to the right, then a 3 over the middle of the next square, and on and on until the last square has a 6 over it. Now, this is a great way to measure things or units. We could measure 6 beans, or 6 spoons, or 6 buttons, but these are 6 unchanging things.

Units are perfect, imaginary things that are used to try and bridge the gap between perfect numbers and imperfect things in the real world. Units are not perfect because of human bias. If you were counting 10 tomatoes (units) just for the hell of it, you would have ten tomatoes (units). If one of the tomatoes was rotten, and you were trying to sell them, then you would really only have 9 tomatoes. As you can see, human bias wreaks havoc with units. Even the old mathematical saying that you can't mix apples and oranges is nonsense. Simply rename them fruits, and the units, thanks to human bias, are changed once again.

This reminds me of another of my Dad's old stories. He and his two little cousins had each been given a cookie by their grandma. While Dad and his cousin were eating their cookies, the third cousin broke her cookie into four pieces and said, "Look, I have four cookies, and you two only have one". She was right, she had four units and they only had one. It hadn't occurred to her that they all had the same amount of cookie, but that's how human bias wrecks units. In the real world there are no permanent, perfect units. Everything appears gradually and disappears gradually, including units. Things accumulate and then *un-accumulate* because of *Change of Physical Position*. Mountains are an accumulation of the earth's crust pushed up into the air, for millions of years, by tectonic pressures, the Andes mountains for instance, and then just as surely un-accumulated by sand borne wind and rain and gravity for millions of years, the Appalachian Mountains, for instance. When we try to measure *Time*, or *Change Of Physical Position*, we run into the problem of *Accumulation and unaccumulation*. This realization is probably why the *Zero* was invented in the first place. Nothing in the real world just appears instantly, it has to accumulate. Even the beans, spoons and buttons mentioned above had to accumulate somewhere. They just seemed to appear instantly when you bought them at the store. Time also accumulates. For instance, let's look at a newborn baby. We won't count pre-birth development here to keep things simpler. A newborn baby is totally helpless, and weighs just a few pounds. One year later it's been completely transformed by accumulation. The timeline we use should be changed to reflect this. Let's go back to our squares diagram to look at this. If we draw a diagonal line from the lower left corner of the first square to the upper right corner of the same square, we can see the rate of accumulation during a year or a century depending on the time span you want to assign to the square. Let's assign 1 year to it for now. If we write the number 50 in the middle of the diagonal line to represent 50% of one year, or 6 months, we can imagine a 6-month old baby, which is already quite a bit different than a newborn after 6 months of accumulation. Now, if we assigned 100 years to our squares, then the 50 would represent 50 years. Any number between 0 and 100 would occur going up along this accumulation line, but they would all be only 2 digit numbers until they reached 100, which would be at the border between the 1st and 2nd squares. If we put a 0 in front of all these double digit numbers in the first square, I think you can see why I think we should start calling the *"First century"*, the *"Zero Century."* Let's go back to our squares again. If we put a 0 in front of the 50 in the middle of the accumulation line (050), the first 0 tells us how many hundreds are in this square, hence the *Zero Century*. In the next square we put the number 150, then the next square 250, and on and on until you put 650 in the last square. Ok, back to the first square on the left. Put a 0 under the far left upright border of the square. Then put a 1 under the next upright border, then a 2 under the next and number all of them up to 6 under the last upright. Now go back to the first square again (bear with me) and write the word *zero* under the 1st square, the word *first* under the 2nd square, the word second under the 3rd, and keep going until you put 6th under the last square.

  WHEW! now that that's over with let's look at the diagram to see what we call these centuries today. Look at the numbers on top of the squares. All the years in the 5th century start with the number 4. All the years in the 6th century start with the number 5. How nutty is that?

  Now look how they could be at the bottom of the diagram. The years in the Zero century all start with the number 0. The years in the 3rd century all start with the number 3, and so on. I think this is the way the timeline ought to be. So, to recap, if we add a *Zero Year* to the current timeline all the millennial years (2000, 3000,

4000) would be the beginning years of each Millennium, and the *Zero Year* would be one year closer to the true birthdate of Christ. Then if we add a *Zero Century* to the current timeline, all the centuries would have the same name and number for example; the 9th Century would be the 900's, instead of today the 9th Century = the 800's.

The years in the 3rd century all start with the number 3, and so on. I think this is the way the timeline ought to be. So, to recap, if we add a *Zero Year* to the current timeline all the millennial years (2000, 3000, 4000) would be the beginning years of each Millennium, and the *Zero Year* would be one year closer to the true birthdate of Christ. Then if we add a *Zero Century* to the current timeline, all the centuries would have the same name and number for example; the 9th Century would be the 900's, instead of today the 9th Century = the 800's. The 1200's would be called the "12th Century", the 1900's would be called the "19th Century", and the 2000's would be called the "20th Century". This would be far easier to remember, because of a *Zero Year* and a *Zero Century*. I think the second thing I would do to our current timeline would be very controversial! People usually don't like big changes to the way they've done things for hundreds of years. But I still think it would be a good idea! After changing the timeline by adding a *Zero Year* for the birth of Christ, I would replace the *Zero Year* with the year 10,000. The year 10,000 would then be Christ's birthday. A number that's pretty majestic and easy to remember, and is also a *Zero Year*. Then, before

110

Diagram # 2   Zero Year / Zero Century

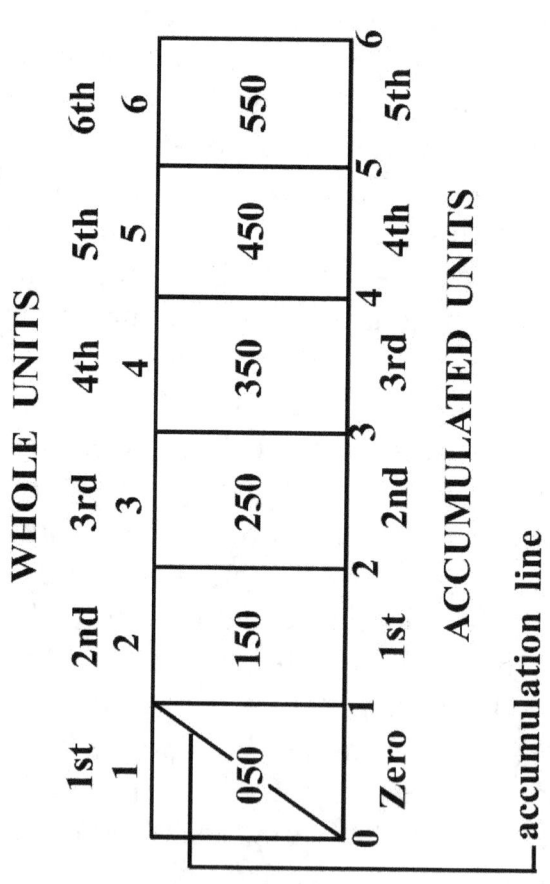

Then, before his birth, the years would continue to get smaller off into the past until they reached the year zero which, like all zeros, would signify nothing. We should also do the same, 10,000 years into the past, at the *Zero Year*. Add another *Pre-Historic Zero Year* and butt it up, nose to nose, against the first *Before Christ Zero Year*. Then the Pre-Historic years would dwindle down to zero at the beginning point of the 10,000-year Before Christ years. Then the years could get bigger again, back into the past from the year zero, which would be 10,000 years before Christ, which was about 4,000 years before the end of the Stone Age. Then nobody but the archeologists would have to remember backward numbers. The historians would hate to have the chore of re-numbering the dates in history, but it would make it a lot easier for the rest of us to remember the past, including future historians.

The years before the zero year could be called P.H. for prehistoric. The years before and after 10,000 wouldn't need to be B.C. or A.D. because all years after the Zero year would be different. You could use A.D. after 10,000 by just removing the 1 from the front of the year to show the number of years since the birth of Christ.

10,000 years before Christ, or 12,000 years ago was a long time ago. It was 4,000 years before the end of the *Stone Age*. Man's only domesticated animal at that time was the dog. Sheep were on the verge of domestication, but all the others were several thousand years in the future. Agriculture began slightly less than 12,000 years ago. The first civilization on earth arose in the Middle East, about 11,000 years ago, and then, writing was about 6,000 years in the future. People were still living in the Stone Age. By putting the year zero back that far in time, it would be far easier to remember what happened when Man rose from the stone age to today's modern civilizations, and keep it all straight in our minds.

Now, let's reverse course and continue into the years after the birth of Christ and on into the present day. How would these changes affect the present calendar? Well, the year 2,013 would become the year 12,013. The year 1946 would become 11,946. This seems like an odd idea at first, but it only adds a "1" to the front of the current date, and all dates back to the birth of Christ. All of our heads are full of dates and years that things happened to us in our lives. To change all those dates drastically would be enormously confusing, and most people would refuse to do it! But if we could devise a system that was extremely simple and easy to remember, we just might make it work, and we would have a much simpler calendar to live with. Let's suppose that 75% of the population (I'm being optimistic) adopted this new way of writing the year. And further, let's suppose the remaining 25% of the population hated this scheme with a passion. "I'll never write that crap!" they'd say. But even though they wrote 2013, everyone could tell instantly which year they were talking about because all they would have to do is add a one to the front of the number. It would also be just as easy for the 25 percenters to remove the first number. You could do this until the year 19,999, which is almost 8,000 years from now. That would give us a fair amount of time to get used to the idea don't you think?

Another worry would be that the new years would be harder to pronounce. I don't think it would be that much of a problem. We don't say "Two-thousand, thirteen" or "One-thousand, nine-hundred, forty-six". We say "Twenty-thirteen" or "Nineteen, forty-six". Instead of pronouncing 12,013 "Twelve-thousand, thirteen", we could pronounce 12,013 "One-twenty-thirteen". Instead of pronouncing 11,946 "Eleven-thousand, nine hundred, forty-six", we could say "One-nineteen-forty-six. But if someone refused to pronounce the year with or without a "1" you would still know which year they were talking about.

This system would not be welcome to many people who would think of it as not commemorating the birth of Christ properly by putting the year one, 10,000 years before his birth. They might say that it would commemorate Mankind's emergence from the Stone Age, and the beginning of civilization instead of Christ's birth where it should be. I would argue that people, their sinful ways, and their need for salvation are the only reason Jesus Christ existed in the first place.
If people were perfect they would have no need for guidance. But they're not perfect, and they do need guidance. Also, most people don't like math, and having to count backwards in time to learn about the past is a huge, mental stumbling block for

almost everyone. People who don't know the past are doomed to repeat it, so I think this new timeline would be an enormous help for people to remember the lessons of the past, and more reliably provide the guidance not to repeat the foolish mistakes of the past again.

CHAPTER 9

PONDERING CALENDAR REFORMERS

I'm proud to say that I invented this 13-month calendar all by myself without help from anyone. I also invented the 10,000-year addition to the timeline all by myself. I had no idea that anyone else had also invented a 13-month calendar and a 10,000 year addition to the timeline until I read about it in a bunch of books I bought to read up on the subject of Time itself. Even when I was a kid, I thought the calendar didn't make any sense. The days were constantly rolling back and forth across the calendar every month, and all the months had different amounts of days. Where is the sense in that? I couldn't see any. When I started thinking seriously about the true nature of Time, I also wondered about the calendar. I saw that each month is about 4 full weeks, give or take a few days. I wondered if you made months of 4 weeks exactly, 28 days only, how many months would fit into a year? I found, using my calculator, that a year would contain exactly 13, 4-week months with only one day left over. I was amazed! There was only one day left over! Why hasn't somebody noticed this before and done something about it?

It did occur to me that since I was a *"genuine math dummy"* that if I could do this, surely someone else with great math skills could have done the same thing a long time ago. After buying the books about Time, I found my suspicions were right. I found there were dozens of calendar reforms proposed over the centuries, most of them totally unworkable. A great example is the French Revolutionary Calendar that was invented during the French revolution that took place at the end of the 1700's. The French revolutionaries despised the Church and the Royalty because they thought that they were plotting together to keep the peasants in poverty, so they wanted to purge any evidence of Christianity or Royalty from everything, including the calendar. So, they changed the names of the months, the weeks, and the days of the entire year. They made each month have 3 weeks of 10 days, and named the months after the types of weather in that month. From Sep 22 to Oct 21 was called Vendemiaire (grape harvest), from Oct 22 to Nov 20, was called Brumaire (mist), for example. The names of all the months (in English) were,

Snowy, Rainy, Windy, Germination, Flower, Meadow, Harvest, Summer Heat, Fruit, Grape Harvest, Mist, and Frost.

An English wit of the day came up with his own names:

Wheezy, Sneezy, Freezy, Slippy, Drippy, Nippy, Showery, Flowery, Bowery, Hoppy, Croppy and Poppy.

Instead of naming each day of the year after saints, which was the custom of the time, they named them after farm produce. Sep 22 was called Grape, Oct 2 was called Potato, Oct 6 was called Donkey, and so on and so on. This extremely complicated calendar lasted for about 12 years till Napoleon came to power and abolished it in Jan 1806.

Another crazy calendar would be popular in Las Vegas. A man named Karl Palmen invented a 13-month calendar that used a deck of cards to name the 13 months from ace to King. Then he named the 4 quarters, in alphabetical order, Clubs, Diamonds, Hearts, and Spades. Then he used the Joker for leap day. These 2 examples are enough to prove my point, but I assure you there are dozens more just as nutty, and often nuttier!

Now I'll describe 4 calendars that I think are the best calendar reforms (almost) to date.

    The Huge Jones Calendar...............1745
    The Marco Mastrofini Calendar.........1834
    The Auguste Compte Calendar.........1849
    The Moses Cotsworth Clendar..........1899

THE GEORGIAN CALENDAR
(not to be confused with the Gregorian calendar)

The first 13-month calendar proposed was by The Reverend Hugh Jones in 1745. Jones was born in Herefordshire, England around 1691. He graduated from the University of Oxford in about

1716, and later emigrated to the U.S., and settled in Virginia at the College of William and Mary, where he taught Mathematics. Jones's most important idea, besides the 13-month calendar was to remove 1 day, 2 days on leap year, from the work week. He wanted to make Leap Day set apart for solemn prayers for King and Country. That sure sounds like a Sunday to me! Also, he wanted to name the new month "Georgy" after King George.

THE MASTROFINI CALENDAR

Italian Abbot Marco Mastrofini's idea for a new perpetual calendar was published in 1834. He said every year should begin on Sunday, Jan 1st, and last 364 days. Then the last day should be set aside and made a holy day. That also sounds like a Sunday to me. Then he said we should set aside Leap day the same way every 4th year. This is also, obviously, a 13-month calendar, but I couldn't find out what he wanted to call the 13th month.

THE AUGUSTE COMPTE CALENDAR

Auguste Compte was a French philosopher and lawyer who was an early socialist. He was an atheist who started his own secular religion he called the Religion of Humanity. Thomas Huxley, the famous English biologist, called his new religion "Catholicism minus Christianity". He proposed his positivist calendar in 1849. It was a

13-month calendar. Compte, true to form, decided to purge any reference to Christianity, or anything else familiar, from his calendar. He renamed all the months for great men of history, and the days after great literary figures. He slipped up a little bit when he renamed January, Moses, and renamed June, St. Paul. He named his 13th month after Francois Bichat, a physician known as the father of modern Histology. For some reason this system didn't catch on! I think he was more interested in stroking his ego than making a workable calendar. Like Jones and Mastrofini, he followed the basic mathematical foundation of the Gregorian calendar.

THE MOSES COTSWORTH CALENDAR

Moses Cotsworth was born in 1859. He was an orphan at 2, and was raised by his Grandparents. They cultivated in him an interest in calendars, charts, and in calculation. He excelled at this and was eventually hired in 1891 to 1895 to do extensive computations to revise the British Railway Rates. While doing this, he became painfully aware of the Crazy-Quilt nature of the days and weeks of the Gregorian calendar. He said we could fix the calendar's problems by making 3 simple changes.

1. Fix all the number dates to the same day names for every month (for example, the first day of every month would be a Sunday the First, then a Monday the second, on through till the last day of the month would be Saturday the 28th. Then change the last week of December to an 8 day week and call the last day, the 29th of December, *Year Day*). This would be repeated forever.

2. Combine the last 13 days of June and the first 15 days of July into a mid year month named Sol.

3. Fix the date of Easter Sunday.

Cotsworth proposed this calendar in 1899, and again in his book "The Rational Almanac", which he published in 1903. Sir Sanford Flemming read Cotsworth's book and was impressed. Flemming built the Canadian Pacific Railway across Canada and also originated the world-wide system of "Standard Time Zones", where every 15 degrees east or west of Greenwich was equal to a one hour change in time. Flemming urged Cotsworth to present his idea to the Royal Society of Canada who were so impressed that they unanimously endorsed the idea. They even reprinted his idea and sent a copy to the British government.

In Feb 1922, advocates of calendar reform selected his idea as the best proposal to reform the calendar. Government, business, and labor leaders all loved the idea. The newly formed League of

Nations loved the idea. However, they forgot to ask the public if they loved the idea. They didn't.

This is a great example of the leaders on top trying to shove one of their favorite ideas down the throats of the people on the bottom. To the average guy in the street, this new calendar was a meaningless nuisance, so they ignored it and the Cotsworth calendar slowly crumbled to dust. A 13-month calendar was, and still is, a great idea, but they should have been selling it to the public, not the government.

THE RICHARD MINIERE CALENDAR

My proposal for a new calendar is only slightly different than the other 13 month calendars presented above. My calendar would follow the Gregorian calendar in all its mathematical precision, just like the others above do. My favorite of the 4 above calendars is the Cotsworth calendar. I think it has the least annoying changes of them all. However, it still has some silly, clumsy things about it that I would streamline.

The silliest thing that Cotsworth did was to stick a new month in the middle of the year, between June and July, and then give it an oddball name (Sol) that doesn't sound anything like a normal month. I propose that the new month not be put at the beginning, or the end, or the middle of the year, but somewhere less noticeable, and also name it something that sounds something like a familiar month name. I propose putting the new month between August and September, and calling it Autember, a blend of the month before and after it. It sounds sort of like Autumn, which is when the new month would occur. Vacation time would be ending, school would be beginning, people would be looking forward to cooler weather and the holidays in the not too distant future. All these distractions would make a new month easier for the public to swallow. I think it would be much less of a poke in the eye than putting Sol between June and July.

The other silly thing Cotsworth did was to put the 365th day in the last week of the year and make the last week of the year 8 days long. Why confuse things like this? I propose that we put the last day of the year as the 29th of December. It would come after Saturday, making it a Sunday. That way the year would begin and end on a Sunday. The year would then have 2 Sundays in a row every year, but they would both be in 2 different years. Then on leap year, the leap day could be put between these 2 Sundays at the end of the year. I don't see why we couldn't make it a Sunday also. The first 2 men to propose a 13-month calendar, Reverend Hugh Jones, and Abbot Marco Mastrofini, both called for these extra days to be set aside as holy days. Sunday is a holy day! I'm just agreeing with the first two men who proposed a 13-month calendar. The preachers might complain about having to work three days in a row instead of their usual one day a week, but it might give them the opportunity to save a few more souls too! I think these two small changes to the Cotsworth calendar would make a new calendar much easier for the public to swallow.

As I said earlier in this chapter, the Cotsworth calendar was being sold to the government not the public. This was a huge mistake!
I can think of several, non-violent examples of this, but I'm sure there are thousands of examples out there. My first example is the "turn right on red" traffic law in Florida. When you come to a full stop at a red light, if you look left and there is no traffic coming from the left, you can make a right turn while the light is still red. I have no idea when this law went into effect, but the public loves it. Even the politicians love it! Any politician in Florida who tries to get rid of this law would be buying a one-way ticket to political oblivion. My second example is "The Miracle of Lourdes", and the "Miracle of Fatima". Whether or not you believe in miracles is irrelevant. These 2 miracles of the Catholic Church are great examples of people leading the way to get what they want in spite of what the government or the church officials want. In both of these miracles, young children say they saw the Virgin Mary several times and she gave them instructions. At first the church and the government thought they were all crazy and thought it was all foolishness. But later, the people were convinced that miracles had occurred, and millions of people began to flock to both places, and the church and the government had little choice but to join the people.

Another example is "citizen band radio" in the United States. Citizen band radio was a 2-way radio introduced in the U.S. as a less technical version of ham radio. It was used for local 2-way communication by the public. You could buy a radio for your house or your car. The radio had 23 channels and you had to fill out a form and mail it in to get a license to operate it. CB radio was not all that well known until a popular song came along, called "Convoy", about truckers that used CB radios to communicate, and a movie, called "Smokey and the Bandit", appeared, which was just a visual version of the song, popularized the radios. This was in the 1980s when there were no cell phones and the popularity of CB radio soared. The government was taken by surprise and wound up having to add more channels, from 23 to 40, and they had to abandon the license rule because most people just bought the radio, used it, and threw away the paperwork that came with it. The people lead the way, and the government followed.

It's a fact that the leaders of people serve at the consent of the people. This might not be so obvious when you look at all the dictators around the world, but it's true! When people get angry enough at their leaders, anything can happen. Also, when the people want something bad enough, they often get it. Most politicians will kiss the public's butt if they think it will keep them in power. But if the public has no interest in a subject, and the politicians don't either then nothing will happen.

A good example of this is what happened on March 21st 1955. Henry Cabot lodge Jr. the

U.S. Representative to the U.N. told the U.N. that since nobody in the U.S. was interested in changing the calendar, and nobody in the Government was interested in changing the calendar either. And furthermore, the U.S. Government thought that any further study of the subject would be a waste of time and money.

If you want to read the official letter see below.

U.S. Opposes Action by U.N. on Calendar Reform
From U.S. Department of State Bulletin, April 11, 1955, p.629

As you can see above, there was no interest in the United States to adopt a new calendar. Obviously, there was a lot of objection on religious grounds, but there was probably a lot of objection to the U.N. too. People in the U.S. just didn't want some foreign power shoving something unwanted down their throats. Another reason, I'll bet, is that most of the public in the U.S. had probably never even heard of this new calendar idea back in 1955. I was alive in 1955, and I never heard anything about a new calendar from my parents, or at school, or from anyone else for that matter. I sure did hear a lot about them not trusting the U.N. though.

Fast-forward to today (2016), and the world is a very different place. School kids, their parents and their grand parents, are walking around with smart phones (hand-held computers) in their pockets. Almost everyone is connected through the internet and the huge volume of information available to everyone is almost unbelievable. Today, the public could easily study this new calendar itself. An app could easily be made to compare the new and the old calendars side by side. This way everyone could judge for themselves the common sense of this idea. And, if the public approved in large numbers, then it would take on a life of its own. This would be change from the bottom up and this slightly new calendar, if popular enough, would have a chance to replace the old calendar to everyone's benefit.

This is a good place to mention a quote by Buckminster Fuller.

"You never change things by fighting the existing reality. To change something, build a new model that makes the existing model obsolete.

CHAPTER 10

LESSONS ABOUT TIME

This last chapter of the book is a good place to go over the lessons we've learned about time. I think the most important lesson we've learned is that time is not an all-encompassing property of the entire universe, but only a very important property of the minds of living beings. We've learned that Time, Minus the Mind, equals Zero, or T-M=0.

I say "only" because we've evicted Time from huge swaths of the universe and deposited it in a few tiny islands of living, universal matter, namely, the minds of living beings. Though the area Time occupies is small, it's of major, major importance to the beings in which it does exist. And since Time is so insular, we have to rethink what we used to think about the non-living physical world around us. We have learned that Time doesn't cause things to happen, but it merely reacts to, and helps us living beings learn from things that happen to us.

We have learned that Time is totally biological. We have learned that Time is our master archivist keeping track of what has happened to us during our lives. At the same time, Time also doubles as an internal navigation mechanism, helping us to travel through life with as little danger and as much advantage to us as possible.

We need to learn that there is no Time in the world around us, only the Time inside us. All of us humans view the world through *Time Colored Glasses*. There is only *existence in a constant state of change* in the non-living universe. The clouds outside your window are the same clouds that Abe Lincoln saw, that Julius Caesar saw, that our cavemen ancestors saw, that the dinosaurs saw. The same water we drink is the same water that ran through the intestines of the dinosaurs and then ran through underground streams for a while, until it resurfaced in a spring and evaporated into the sky again and again never stopping. This is the constant change that occurs in all physical matter all by itself without any help from Time at all.

We learned that a slightly new calendar and timeline would be of great benefit to humanity, but it would have to be sold first to the public, and if the public was convinced it was a good idea then governments would fall in behind it and support it.

We also learned what a clock is. Before we knew what Time was we couldn't describe what a clock was because we had no idea what a clock measured. But now we know that a clock is a machine that measures *Change Of Physical Position*, or *COPP*. A watch or clock is a Copp meter, or Coppometer.

Another good lesson is about the mistaken feeling that anything from the past is old, obsolete, and no good anymore. It's worn out, behind the times, not cool, from the stone-age, funny looking, obsolete, and only of interest to old geezers. This is usually the viewpoint of young people who are older teenagers or early twenty-somethings that have recently found that Mom and Dad are actually frail human beings, instead of the large people that know everything. Then, after being slapped around for 30 or 40 years by kids, spouses, bosses, and life in general, they begin to see that Mom and Dad weren't quite as dumb as they had previously thought. About this time, they hopefully begin to see that Time has nothing to do with whether an idea, or a thing, is good or bad.

A good example of an old thing that's just as good as ever is the over 3,000-year old gold and blue funeral mask of King Tut. This thing is gorgeous, and always will be. King Tut's funeral mask is as beautiful today as when it was made, and will be for thousands of years. The old saying that "a thing of beauty is a joy forever" applies here. We can say that its beauty is Timeless. And since it's a non-living thing, it is Timeless. And yet it's also an idea, set in gold metal, blue glass and gemstones, sent thousands of years into the future by the ancient Egyptians to display their reverence for their boy king.

A good example of an old idea that's just as good as ever is the United States Constitution. It's given the people of the United States over two hundred years of un-paralleled peace and prosperity, and will give centuries more of the same if the citizens of the United States can keep it's Constitution intact. When Ben Franklin left

Independence Hall at the close of the constitutional convention of 1787, a lady on the street asked him "Well, what have we got, a republic or a monarchy?" Franklin replied "A Republic...if you can keep it." This is a truly Timeless statement that applies to anyone, anywhere that tries to keep a true Republic. Only Time, or Change Of Physical Position, will tell.

We think of certain things as being obsolete because something newer has replaced them. For instance, that sailing ships are obsolete. That their time has come and gone. Gasoline and diesel fuel have replaced them. But, although very large ships are fueled, the sailboat hasn't gone away. I'm not talking only about Pleasure boats, there are still a lot of working sailboats left in the world in poorer countries around the world, simply because they work. When gas prices soar at different times, even large ocean going oil tankers are toying with the idea of sails on those gigantic ships. A company called SkySails based in Hamburg Germany wants to turn cargo ships into giant wind surfers. They have built a 1700 Sq. Ft. kite that can generate as much pull as the thrust of an Airbus A318 jet engine. They claim these sails can cut fuel consumption as much as 50%.

The old saying "Everything old is new again" comes to mind.

Another area where the term obsolete rears its ugly head is in the military. War machines seem to become obsolete every 10 years or so, even faster during a war. However, this is not completely true. The helmet was used by the ancient Greeks to protect their heads in battle, and though they've changed shape over the years, the helmet is still in use today in the most modern of militaries around the world. Helmets use no fuel, and they simply work. Other supposedly obsolete weapons were used during the Vietnam War such as the bow and arrow and the crossbow. They also don't need fuel, they're silent, and they simply work. The knife and bayonet are as ancient as warfare, but they are still used in every military even today. Let's not forget the most ancient weapon of all, the common foot soldier. You may have airplanes and ships and tanks, but the foot soldier is the one who wins the war, and always will be. So why are these old-fashioned weapons still being used, they're obsolete aren't they? The answer is no, because they still work. A beefy arm connected to a hand with a powerful grip on a rock the size of a brick can bash in somebody's head as easily today as it could in the hand of a caveman 70,000 years ago. It's primitive, but it still works.

Does Time cause things to become obsolete? Time does play a role in obsolescence because Time is a biological mechanism that helps us learn from past failure or success. We review our failures and try to use our imaginations to tweak our future activities to try and turn our past failure into a future success. However, *Change Of Physical Position* also plays a role. Copp is always in motion in the non-living world changing conditions under which people live. Floods, earthquakes, severe weather that causes crop failures, combined with human failures such as economic crashes, disease, starvation and invasion, or threat of invasion by other people, will often cause people to go to war with one another. This war will cause each side to try and innovate a new weapon or way of doing things, such as spying, to make the other sides weapons, or way of doing things, obsolete.

Another area where obsolescence is rampant is in the fashion industry. A whole encyclopedia could be written about this. The vast majority of this type of obsolescence, of course, is in women's fashion. Women love clothes! Women love new clothes even more! To most women clothes are as much a form of entertainment as a necessity, men...well, not so much. There are men who are fashionable and buy a lot of clothes, but most men buy and wear just enough clothes just to keep warm or cool, or from being arrested for indecent exposure.

So, women's clothes are where the big money is. This is one more area where the old saying "Everything old is new again" really applies. Women who want new clothes, and stores who want to sell them new clothes, get together several times each year to make each other's dreams come true! But eventually, the fashion business runs out of new ideas and they have to bring back old styles and claim that the old styles are "New" again, and of course to younger women they are new.

A good example of this is my Mom's old shoes in the attic when I was about ten in 1956.

She showed me a pair of bright red, alligator platform shoes she had worn back in the 1940's. They looked almost new to me and I asked her "Mom, how come you don't wear these anymore?" "Oh, I can't wear those anymore they're out of style." "Oh," I said. "Mom, how come the bottoms are so thick?" I said? "Oh, that's just the style", she said. I thought to myself, she could wear those every day for the next 50 years before the soles wore out because they were so thick. Then she showed me a pair of shoes that looked like men's shoes called saddle oxfords. Saddle oxfords do look like men's shoes made out of white leather, but the middle part where the laces go are made of brown leather. "How come you wore men's shoes Mom", I asked? "They're not men's shoes, they're women's shoes, I wore them." "Oh", I said. "How come you don't wear them anymore, I asked?" "I can't wear them anymore because they're out of style", she said. "Mom, how do you know if something is in style or not", I asked. "I just look at what other women are wearing, if they're wearing it, it's in style. If they're not wearing it, it's not in style." Mom seemed to be right because I never saw any shoes like the ones she had in the attic on women walking around in public.

Fast-forward about 12 years and I'm in the U.S. Air Force stationed at Moody Air Force Base in Valdosta, Georgia. Valdosta is a small town in southern Georgia that's about a hundred miles from any big city. There was one High School there, Valdosta High School, home of the Wildcats football team. High School Football was a BIG

DEAL in Valdosta. Most of the town showed up to watch the Wildcats play. When I first went into town in Valdosta with some of the guys on base, we all noticed that almost all the teenaged girls were wearing saddle oxford shoes. Their shoes weren't old either, they were brand new. I mentioned this to my Mom during a phone call home one day and she was shocked! "Those things went out of style 20 years ago", she said. Then I said, "but Mom, you're not a teenaged girl in Valdosta Georgia, and like you told me years ago, if all the women are wearing it, it's in style."

So, we can see that when it comes to fashion, timing is very important when it comes to introducing young women to old fashions presented as something "New", but Time itself has nothing to do with the changing of fashions or anything else in the real world. The only thing Time can change is our opinion, and our opinion of "What's in style".

There is no better example of planned obsolescence than the automotive industry in the United States during the 1950s. I observed this phenomenon in person as a kid in elementary and junior high school. Every boy my age knew every model of car by heart. We knew which one had the widest, longest, or highest tail fins, and to which model they belonged. The first car with tail fins was the 1949 Cadillac. They were pretty small really, but every year they got bigger until in 1959 they were huge fins about 2 feet tall with 2 bullet shaped tail-lights on each fin. When my friends and I first saw them, we all gasped a collective

WOW!! Every year, after the new models came out, it was painfully obvious to everyone that you either had a New car or that your current car was now OLD!

By the time 1960 rolled around the tail fins, and the tail fin craze began to fade away, but it sure was fun while it lasted.

Now, let's talk about old, obsolete things that people actually want! Instead of old, dusty, out of style furniture, they're rebranded as "Antiques". Instead of old clothes your grandma wore when she was young, they're rebranded "Vintage clothing".

One of the most amazing lessons about Time is that there is no difference between today and the day Christ was born. There is no difference between today and the first day that work was begun on the Pyramids in ancient Egypt. There is no difference between today and the day that the pilgrims landed on Plymouth Rock. There is also no difference between *any* 2 dates you can think of in the real, physical world when we're talking about Time, since there is no Time in the real world. But in the real world, there is always an enormous amount of physical change taking place. It's a basic property of all matter that exists. Matter has mass and it is always in constant motion, or to be very specific, *Change Of Physical Position*, or *COPP*. Even a change as concrete feeling as day and night is only an effect caused by the earth constantly spinning with sun shining on it from one direction. It doesn't change the earth in any way by its spinning, except maybe by a slight temperature change on the sunlit side. The earth in its orbit is the same way. It's just constantly spinning on its orbit around the sun. There really is no beginning or end of the year, there is only constant spinning in its orbit. We humans here on earth notice the year because of the change of seasons, and later we noticed the positions of the stars at night continuously moving westward and returning to the same star patterns every year. But the Time in our heads has nothing to do with the change, Time only helps us notice it, and then Time has done the job it was designed to do.

Now that we know that Time only exists in our heads, and that Change Of Physical Position is what we always thought time to be, will we stop using the word time? Will we start asking people " Hey buddy, can you tell me what change of physical position it is?" Well, maybe if you're a scientist working on an atomic clock or something, but for everybody else the answer is no! Nobody wants to complicate things, they'll always ask "What time is it?" That's because Time is a property of being human, and it always will be, because it works so well for us. Also, most people will probably never even read this book, and have assumed that scientists have known what Time is for centuries anyway, so for most people, the subject will never come up.

Now, for our final lesson. As I said in the first chapter of this book, there is no past, present, or future in the non-living universe, only existence, in a constant state of change.

Think of this statement as you read the beginning of the first chapter of Ecclesiastes in the King James Version of the Bible:

1. The words of the Preacher, the son of David, King in Jerusalem.

2. Vanity of vanities, saith the Preacher, vanity of vanities; all is vanity.

3. What profit hath a man of all his labour which he taketh under the sun?

4. One generation passeth away, and another generation cometh: but the earth abideth forever.

5. The sun also ariseth, and the sun goeth down, and hasteth to his place where he arose.

6. The wind goeth toward the south, and turneth about unto the north; It whirleth about continually, and the wind returneth again according to his circuits.

7. All the rivers run into the sea; yet the sea is not full; unto the place from whence the rivers come, thither they return again.

8. All things are full of labour; man cannot utter it: the eye is not satisfied with seeing, nor the ear filled with hearing.

continued on next page

9. The thing that hath been, it is that which shall be; and that which is done is that which shall be done: and there is no new thing under the sun.

10. Is there anything whereof it may be said, See, this is new? It hath been already of old time, which was before us.

11. There is no remembrance of former things; neither shall there be any remembrance of things that are to come with those that shall come after.

Books About Time I Have Read

Since I hate Indexes but need to cite the work of others, I decided to list their books instead. I got something from all of them, more from some, and less from others. What I got from all of them was the enormous confusion we all feel about what Time really is.

TIME, IT'S ORIGIN, IT'S ENIGMA, IT'S HISTORY
    By Alexander Waugh

THE END OF TIME
    By Julian Barbour

TO INFINITY AND BEYOND
    By Eli Maor

ABOUT TIME
    BY Paul Davies

CALENDAR
    By David Ewing Duncan

HYPERSPACE
    By Michio Kaku

MAPPING TIME
    By E.G. Richards

Continued next page

TIME'S ARROW AND ARCHIMEDE'S POINT
By Huw Price

THE OXFORD COMPANION TO THE YEAR
By Bonnie Blackburn and Leofranc Holford-Strevens

e: THE STORY OF A NUMBER
By Eli Maor

LONGITUDE
By Dava Sobel

AZIMOV'S BIOGRAPHICAL ENCYCLOPEDIA OF SCIENCE AND TECHNOLOGY
By Isaac Azimov

WIKIPEDIA,
By Everybody

www.ingramcontent.com/pod-product-compliance
Lightning Source LLC
Chambersburg PA
CBHW052211220526
45471CB00004B/1913